THINKING
BETTER

THINKING BETTER

BETTER

THE ART OF THE SHORTCUT
IN MATH AND LIFE

MARCUS DU SAUTOY

BASIC BOOKS
NEW YORK

Basic Books
Hachette Book Group
1290 Avenue of the Americas, New York, NY 10104
www.basicbooks.com

Printed in the United States of America

First Edition: October 2021

Published by Basic Books, an imprint of Perseus Books, LLC, a subsidiary of Hachette Book Group, Inc. The Basic Books name and logo is a trademark of the Hachette Book Group.

The Hachette Speakers Bureau provides a wide range of authors for speaking events. To find out more, go to www.hachettespeakersbureau.com or call (866) 376-6591.

The publisher is not responsible for websites (or their content) that are not owned by the publisher.

Library of Congress Cataloging-in-Publication Data
Names: Du Sautoy, Marcus, author.
Title: Thinking better : the art of the shortcut in math and life / Marcus du Sautoy.
Description: First edition. | New York: Basic Books, 2021. | Includes index.
Identifiers: LCCN 2021008118 | ISBN 9781541600362 (hardcover) | ISBN 9781541600379 (ebook)
Subjects: LCSH: Mathematics. | Mathematics—Methodology. | AMS: General and overarching topics; collections—General and miscellaneous specific topics—Popularization of mathematics. | General and overarching topics; collections—General and miscellaneous specific topics—Methodology of mathematics. | General and overarching topics; collections—General and miscellaneous specific topics—General applied mathematics. | General and overarching topics; collections—General and miscellaneous specific topics—Mathematics for nonmathematicians (engineering, social sciences, etc.).
Classification: LCC QA39 .D77 2021 | DDC 510—dc23
LC record available at https://lccn.loc.gov/2021008118

ISBNs: 978-1-5416-0036-2 (hardcover), 978-1-5416-0037-9 (ebook)

LSC-C

Printing 1, 2021

TO ALL MATHEMATICS TEACHERS, BUT ESPECIALLY
MR. BAILSON, WHO SHOWED ME MY FIRST
MATHEMATICAL SHORTCUT

CONTENTS

DEPARTURE

YOU HAVE A CHOICE. The obvious path is a long slog, with no beautiful vistas on the way. It is going to take you forever and sap all your energy, but it will eventually get you to your destination. There is a second path, however. You've got to be sharp to spot it veering off the main path, seemingly taking you away from your destination. But you spot a signpost that says SHORTCUT. It promises a quicker off-road route that will get you to your destination faster and with minimal energy expenditure. There might even be the chance of a stunning view on the way. It's just that you are going to have to keep your wits about you to navigate this path. It's your choice.

This book is pointing you toward that second path. It's your shortcut to the better thinking you'll need to negotiate this unorthodox route and get you to where you want to go.

It was the lure of the shortcut that made me want to become a mathematician. As a rather lazy teenager, I was always looking for the most efficient path to my destination. It was not that I wanted to cut corners; I just wanted to achieve my goal with as little effort as possible.

So when my mathematics teacher revealed to me at the age of twelve that the subject we were learning in school was really a celebration of the shortcut, my ears pricked up. It started with a simple story featuring a nine-year-old boy named Carl Friedrich Gauss. Our teacher transported us back to 1786 in the town of Brunswick, near Hanover, where the young Gauss grew up. It was a small town, and the local school only had one teacher, Herr Büttner, who had to somehow teach the town's one hundred children in just one classroom.

My own teacher, Mr. Bailson, was a rather dour Scot who kept strict discipline, but it sounded like he was a softy compared to Herr Büttner. Gauss's teacher would stride up and down the benches brandishing a cane to maintain discipline among the rowdy class. The classroom itself, which I've subsequently visited on a recent mathematical pilgrimage, was a drab room with a low ceiling, little light, and uneven floors. It felt like a medieval prison, and Büttner's regime sounded as if it matched the setting.

The story goes that during one arithmetic lesson Büttner decided to set the class a rather tedious task that would occupy them long enough so that he could take a nap. "Class, I want you to add up the numbers from 1 to 100 on your slates. As soon as you are done, bring your slates to the front of the class and place them on my desk."

Before he'd even finished the sentence, Gauss was on his feet and had placed his slate on the desk, declaring in Low German, "Ligget se"—there it is. The teacher looked at the boy, shocked at his impertinence. The hand holding the cane quivered with anticipation, but he decided to wait until all the students had submitted their slates for inspection before upbraiding the young Gauss. Eventually the class had finished and Büttner's desk was a tower of slates covered in chalk and calculations. Büttner began to work his way through the pile, starting with the last slate placed on the top. Most of the calculations were wrong, the students having invariably made some arithmetic slip on the way.

Eventually he arrived at Gauss's slate. He began preparing his rant at the young upstart, but when he turned over the slate, there was the correct answer: 5050. And there were no extraneous calculations. Büttner was shocked. How had this nine-year-old found the answer so quickly?

The story goes that the precocious young student had spotted a shortcut that helped him avoid the hard work of actually doing much arithmetic. What he had realized was that if you add up the numbers in pairs:

$$1 + 100$$
$$2 + 99$$
$$3 + 98$$
$$\cdots$$

then the answer for each pair was always 101. But there were 50 pairs. Hence the answer was

$$50 \times 101 = 5050$$

I remember being electrified by this story. To see Gauss's insight into how to shortcut all this horribly tedious and labor-intensive work was a revelation.

Although the story of Gauss's schoolroom shortcut is probably more legend than fact, it nonetheless captures beautifully an important point: mathematics involves not tedious calculation, as so many think, but rather strategic thinking.

"That, my dear students, is mathematics," my teacher announced at the end of the Gauss story. "The art of the shortcut."

Hello, my twelve-year-old self thought. *Tell me more!*

Getting Further Faster

Humans use shortcuts all the time. We have to. We have a short amount of time to make a decision. We have limited mental capacity to navigate complex problems. One of the first strategies that humans developed as a pathway to solving complicated challenges is the idea of heuristics—the process by which we make problems less complex by ignoring, either consciously or unconsciously, some of the information that's coming into the brain.

The trouble is that most heuristics that humans use lead to bad judgments and biased decisions, and generally aren't good enough to do the job. We might know one thing from experience and then try to extrapolate to all other problems by comparing them to this one thing we know. We judge the global by the knowledge of the local. This was fine when our environment didn't extend too much beyond the small region of savannah we inhabited. But as our neighborhood expanded, these heuristics didn't give us good ways of understanding how things worked beyond our local knowledge. This is the moment we had to start to develop better shortcuts. Those tools are what we today call mathematics.

To find good shortcuts requires the ability to lift yourself out of the geography you're trying to traverse. If you are in the landscape, then often you can rely only on what you see immediately around you. Although each step feels like it is taking you in the right direction, the cumulative result of the path might take you the long way around to your destination or lead you astray completely. That's why humans developed a better way of thinking: the ability to lift yourself out of the minutiae of the task at hand and understand that there might be an unexpected path that could get you to your destination faster and more efficiently.

This is what Gauss did with the challenge his teacher set the class. While the other students starting plodding from one number to the next, adding each new number they encountered to the tally, Gauss surveyed the problem in its totality, understanding how to use the beginning and end of the journey to his advantage.

Mathematics is all about this ability to use higher-level thinking to see structure where before we just saw random meandering pathways, to mentally lift yourself out of the landscape and look down from a great height to see the true lay of the land. Shortcuts emerge when problems are mapped in this way. And once we started to exploit the ability to see structure in our mind's eye without physically encountering it, this capacity for abstract thinking unleashed human civilization's extraordinary advances over the centuries.

The journey to better thinking began five thousand years ago around the Nile and the Euphrates. Humans wanted to find cleverer ways of building the city-states that were blossoming alongside these rivers. How many blocks of stone would be needed to build a pyramid? What area of land needs to be cultivated with crops to feed a city? What changes in river height were indicators of a forthcoming flood? Those who had the tools to find shortcuts to solving these challenges were those who rose to prominence in these emerging societies. The success of mathematics as a shortcut to the rapid development of these civilizations launched the subject as a powerful tool for those wishing to get further faster.

Time and again a gear change in civilization was effected by the discovery of new mathematics. The explosion of mathematics during the Renaissance and beyond, which gave us tools such as calculus, offered scientists extraordinary shortcuts to efficient engineering solutions. And today mathematics is behind all the algorithms being implemented on our computers to assist us through the modern digital jungle, offering shortcuts that help us find the best routes to our destinations, the best websites for our internet searches, and even the best partners for a journey through life.

It is interesting to note, however, that humans weren't the first to exploit the power of mathematics to access the best way to tackle a challenge. Nature has been using mathematical shortcuts to solve problems long before we arrived. Many of the laws of physics are based on Nature always finding a shortcut. Light travels along the path that gets it to its destination fastest, even if that involves bending around a large object like the sun. Soap films create the shapes that cost the lowest amount of energy—the bubble makes a sphere because this symmetrical shape is the one with the smallest surface area and therefore costs the minimum energy. Bees make hexagonal cells in their hives because the hexagon uses the least amount of wax to contain a fixed area. Our bodies have found the most energy-efficient way of walking to transport us from A to B.

Nature is lazy, like humans, and wants to find the lowest-energy solutions. As the eighteenth-century mathematician Pierre-Louis

Maupertuis wrote: "Nature is thrifty in all its actions." It is extremely good at sniffing out shortcuts. Invariably it has a mathematical rationale to it. And often the discoveries of shortcuts by humans materialize out of our observations of how Nature solves a problem.

The Journey Ahead

In this book I want to share with you the arsenal of shortcuts that mathematicians such as Gauss have developed over the centuries. Each chapter will introduce a different sort of shortcut with its own particular flavor. But all of them have the aim of transforming you from someone who has to slog through the hard work of solving a problem to someone who can hand in their slate with the answer before everyone else.

I have chosen to take Gauss as a companion on our journey. His classroom success launched him on a career that marks him out for me as the prince of the shortcut. Indeed, the plethora of breakthroughs he made during his lifetime span many of the different shortcuts that I will introduce throughout the book.

By telling the stories of the shortcuts that mathematicians have amassed over the centuries, I hope this book will act as a tool kit for all those who want to save time doing one thing so that they can spend more time doing something more exciting. Very often these shortcuts are transferable to problems that don't at first glance seem mathematical in nature. Mathematics is a mindset for navigating a complex world and finding the pathway to the other side.

This is why mathematics really deserves to be a core subject in the educational curriculum. Not because it is absolutely essential that we all know how to solve a quadratic equation; frankly, when has anyone ever needed to know that? The essential skill is understanding the power that algebra and algorithms play in solving such a problem.

I begin the journey to better thinking with one of the most powerful shortcuts mathematicians have developed: patterns. A pattern is often the best sort of shortcut. Spot the pattern and you've found the shortcut to continuing the data into the future. This ability to spot an underlying rule is the basis of mathematical modeling.

Quite often the role of the shortcut is to understand the foundational principle that unites a whole slew of seemingly unrelated problems. The beauty of Gauss's shortcut is that even if the teacher tries to make it harder by asking you to add the numbers up to 1000 or 1,000,000, the shortcut still works. While adding numbers up one by one would get increasingly time-consuming, Gauss's trick is unaffected by an increase in the number of numbers you're adding up. To add the numbers up to 1,000,000, just pair them up again (1 + 1,000,000, 2 + 999,999 . . .) to get 500,000 pairs that each add up to 1,000,001. Multiply these two together (500,000 × 1,000,001) and bingo: you've got your answer. The tunnel that provides a shortcut through the mountain is unaffected by the mountain getting taller.

The power of creating and changing language turns out to be a very effective shortcut. Algebra helps us recognize the underlying principles behind a whole range of different-looking problems. The language of coordinates turns geometry into numbers and often reveals shortcuts that were not visible in the geometric setting. Creating language is an amazing tool for understanding. I remember wrestling with an extraordinarily complex setup that needed many conditions to pin down. My doctoral supervisor's suggestion that I "give it a name" was a revelation—it truly allowed me to shortcut thought.

Whenever I mention the idea of the shortcut, invariably people think I am trying to cheat somehow. The word "cut" sounds like you could be cutting corners, so it's important right from the outset to distinguish between shortcuts and cutting corners. I'm interested in the clever path to get to the correct solution. I'm not interested in finding some shoddy approximation to the answer. I want complete understanding, but without unnecessary hard work.

That said, some shortcuts are about approximations that are good enough to solve the problem at hand. In some sense, language itself is a shortcut. The word "chair" is a shortcut to a whole host of different sorts of things we can sit on. But it is not efficient to come up with a different word for every distinct instance of an example of a chair. Language is a very clever low-dimensional representation of the world around us that allows us to efficiently communicate to others and facilitates our path through the multifaceted world we live in. Without the shortcut of single words for multiple instances, we would be overwhelmed by noise.

In mathematics too I will reveal how throwing away information is often essential to finding a shortcut. The idea of topology is geometry without measurement. If you are on the London Underground, a map showing how stations are connected is much more useful for finding your way around London than a geometrically accurate map. Diagrams are also a powerful shortcut. Again, the best diagrams discard anything that is extraneous to navigating the problem at hand. But as I shall illustrate, there's often a fine line between a good shortcut and the dangers of cutting corners.

Calculus is one of humans' greatest inventions for finding shortcuts. Many engineers depend on this bit of mathematical magic to find the optimal solution to an engineering challenge. Probability and statistics have been a shortcut to knowing a lot about a huge data set. And mathematics can often help you find the most efficient path through a complex geometry or tangled network. One of the staggering revelations I had as I fell in love with mathematics was its ability to find shortcuts to navigate even the infinite—a shortcut to get from one end of an infinite path to the other.

Each chapter begins not with an epigram but a puzzle. Often these puzzles involve a choice: the long slog or, if you can find it, the shortcut. Each puzzle has a solution that takes advantage of the shortcut that is at the heart of that puzzle's chapter. They are worth tackling before you read the chapter, as often the more time you spend battling to get to your destination the more you appreciate the shortcut when it is finally revealed.

What I have discovered on my own journey is that there are different sorts of shortcut. Because of this, I spend time highlighting the multiple approaches you might take to the journey you are about to embark on, and show that you will get to your destination faster by using the most effective shortcut. There are shortcuts that are already waiting there in the terrain for us to take advantage of them; it's just that you might need a signpost to point you in the right direction or a map to show you the way. There are shortcuts that won't exist if you don't do a lot of hard work to carve them out—like the tunnel that takes years to dig but once there allows everyone else to follow you through to the other side. There are shortcuts that require totally escaping the space you are in—the wormhole from one side of the universe to the other, or the extra dimension that shows how two things are much closer than you imagine provided you can step out of the confines of the current world. There are shortcuts that speed things up, shortcuts that cut down the distance you need to travel, and shortcuts that reduce the amount of energy you need to expend. Somewhere there is a saving that is worth the time to find the shortcut.

But I've also recognized that there are times when the shortcut misses the point. Maybe you want to take your time. Maybe the journey is the thing. Maybe you want to expend energy in an attempt to lose weight. Why go on a walk in Nature for the day if you curtail the pleasure of the walk by taking a shortcut home? Why read a novel rather than a synopsis on Wikipedia? But it's still good to know you've got the option of a shortcut even if you decide to ignore it.

The shortcut is to some extent about our relationship to time. What do you want to spend your time doing? Sometimes it is important to experience something in time and there is little value to finding a shortcut that cheats you of the feeling. Listening to a piece of music can't be shortcutted. It takes time. But on other occasions life is too brief to spend time getting to where you want to be. A film can condense a life into ninety minutes; you don't want to witness every action of the character you are following. Taking a flight to the other side of the world is a shortcut to walking there and means you can begin your vacation sooner; if you could shorten the flight even further, you probably would. But there are times when people want to experience the slow version of getting to their destination. Pilgrimage abhors the shortcut, for instance. And I never watch film trailers, because they shortcut the film too much. But it is still worth having the choice.

Shortcuts in literature are invariably paths that lead to disaster. Little Red Riding Hood never would have met the wolf if she hadn't strayed from the path in search of a shortcut through the wood. In Bunyan's *Pilgrim's Progress*, those who take a shortcut around Difficulty Hill get lost and perish. In *The Lord of the Rings* Pippin warns that "shortcuts make long delays" (although Frodo counters that inns make even longer ones). Homer Simpson swears after his disastrous detour on the way to Itchy and Scratchy Land, "Let us never speak of the shortcut again." The dangers inherent in taking shortcuts are well summed up in the film *Road Trip*: "Of course it's difficult—it's a shortcut. If it was easy, it would just be 'the way.'" This book looks to rescue the idea of a shortcut from these literary tropes. Rather than the road to disaster, the shortcut is the road to freedom.

Human Versus Machine

One thing that sparked my desire to write this book celebrating the art of the shortcut is the increasingly common perception that the human race is about to be superseded by a new species that doesn't need to bother with shortcuts. We are now living in a world where computers are able to do more computations in an afternoon than I can do in a lifetime. Computers can analyze all of world literature in the time that it takes me to read one novel. Computers are able to analyze many more variations of a game of chess than the few moves I can hold in my head. Computers can explore the contours and pathways covering planet Earth faster than it takes me to walk to the corner store.

Would a computer today come up with Gauss's shortcut? Why would it bother, when it can add up the numbers from 1 to 100 in the nth of an nth of a blink of an eye?

What hope is there for humankind to keep up with the extraordinary speed and nearly infinite memory of our silicon neighbors? The computer in the film *Her* declares to its human owner that the pace of human interaction is so slow, it prefers to spend time with other operating systems that can match its speed of thought. The computer looks at the speed at which humans operate the way we look at the slow pace at which a mountain emerges and erodes.

But maybe there is something that gives humans the edge. The limitations on the human brain's ability to perform computations simultaneously compared to a computer's and the physical shortcomings of the human body compared to the strength of a mechanical robot will force humans to stop and think whether there is a way to avoid all the steps that a computer or robot finds trivial.

Faced with a seemingly unassailable mountain, humans will instead seek out the shortcut. Rather than trying to go over the top of the mountain, is there perhaps a sneaky way around? And often it is the shortcut that leads to a truly innovative way to solve a problem. While the computer plows on, flexing its digital muscles, the human sneaks to the finish line, thanks to cunning shortcuts that help us avoid all the hard work.

Slackers, take note. I think laziness is our saving grace, what will protect us against the onslaught of the machine. Human laziness is a really important part of finding good new ways to do things. When a computer is faced with a problem, we know what it will say: *Well, I've got these computational tools, so I can just bash my way through the problem.* But I often look at a problem and think, *This is just getting too complicated—let me try to step back and figure out a shortcut.* Because a computer doesn't get tired and it's not going to be lazy, maybe it will miss things that our laziness takes us to. Because we don't have the ability to plow right through problems the way a computer would, we're forced to find clever ways to handle them.

There are many stories of how innovation and progress grew out of laziness and a desire to avoid hard work. Scientific discovery often emerges from a mind left idling. The chemist August Kekulé is said to have come up with the ring structure of the benzene molecule after falling asleep and dreaming of a snake swallowing its own tail. The great Indian mathematician Srinivasa Ramanujan often spoke of his family goddess, Namagiri, writing equations in his dreams. As he recalled: "I became all attention. That hand wrote a number of elliptic integrals. They stuck to my mind. As soon as I woke up, I committed them to writing." Jack Welch, as chairman and CEO of General Electric, spent an hour each day in what he called "looking out the window time." A new invention is often born of someone who can't be bothered to do things the hard way.

Laziness doesn't mean that you do nothing. And this is a really important point. Finding shortcuts often requires hard work. It's something of a paradox. Although the motivation to find a shortcut might come from the desire to avoid boring work, or perhaps to cope with the boredom that idleness brings, the result is often intense periods of deep thought. There is a fine line between idleness and boredom, and it is this that often is the catalyst for the hunt for a shortcut that can then involve a great amount of labor. As Oscar Wilde wrote, "To do nothing at all is the most difficult thing in the world, the most difficult and the most intellectual."

Doing nothing is often a precursor to great mental progress. A paper published in 2012 in *Perspectives on Psychological Science* entitled "Rest Is Not Idleness" reveals how important our default mode of neural processing is to cognitive ability. This mode is often suppressed when our attention is too focused on the outside world. The recent surge of interest in mindfulness suggests the value of stilling the mind as a pathway to enlightenment. Often it means you prefer to play rather than work. But play is often the place to foster creativity and new ideas. It is one of the reasons that the offices of start-ups and math departments often contain pool tables and board games as well as desks and computers.

Perhaps society's disapproval of laziness is a way of controlling and curtailing those who prefer not to conform. The real reason the lazy person is regarded with suspicion is that laziness is the mark of someone not prepared to play by the rules of the game. Gauss's teacher saw his pupil's shortcut to doing hard work as a threat to his authority.

Idleness has not always been shunned. Samuel Johnson very eloquently argued in favor of laziness: "The Idler . . . not only escapes labours which are often fruitless, but sometimes succeeds better than those who despise all that is within their reach." As Agatha Christie admitted in her autobiography, "Invention, in my opinion, arises directly from idleness, possibly also from laziness. To save oneself trouble." Babe Ruth, one of the best home-run hitters baseball has ever seen, apparently was motivated to hit the ball out of the stadium because he hated having to run between bases; when he hit a homer, he could take his time rounding the bases.

Choosing to Work

I do not wish to imply that all work is bad. Indeed many people get great value out of the work they do. It defines their identity. It gives them purpose. But the quality of the work is important. Generally, the work we find valuable is not a series of tedious, mindless tasks. Aristotle distinguished between two different sorts of work: *praxis*, which is action done for its own sake, and *poiesis*, or activity aimed at the production of something useful. We are happy to look for shortcuts in the second sort of work, but there seems little point in chasing the shortcut if the pleasure is in doing the work for its own sake. Most work seems to fall into the second category. But surely the ideal is to aspire to work of the first kind. That is where the shortcut aims to take you. The shortcut is not about eliminating work; it wants to lead you on a path to meaningful work.

The principle behind the new political movement Fully Automated Luxury Communism is that with advances in AI and robotics, machines can take over our menial work, leaving time for us to indulge in work we find meaningful. Work becomes a luxury. The cultivation of good shortcuts should be added to the list of technologies steering us toward a future of work that is undertaken for the joy of it rather than as a means to an end. This was Marx's aim with communism: to remove the difference between leisure and work. "In a higher phase of communist society . . . labour has become not only a means of life but life's prime want." The shortcuts we have created promise to take us away from what Marx called the "realm of necessity" and lead us instead into the "realm of freedom."

But there are some places where you can't get away from hard work. How can a lazy person learn a musical instrument? Write a novel? Climb Everest? Even here, though, I shall illustrate how shortcuts can help you maximize the value of the hours you put in at your desk or in training. The book is punctuated by conversations I've had with high achievers to see whether shortcuts are possible in their professions or if you just can't avoid the ten thousand hours of practice that Malcolm Gladwell says are necessary to get to the top of your profession. I've been intrigued to find out whether the shortcuts that people have found resonate with those I've learned in mathematics, or whether there might be new sorts of shortcuts that I've not been aware of but which might prompt new modes of thinking in my own work. But I'm also fascinated by those challenges where no shortcuts are possible. What is it about certain domains of human activity that preclude the power of the shortcut? Time and again, it turns out, the human body is often the limiting factor. To change or train or push the human body to do new things quite often takes time and repetition, and there are no shortcuts to speed up those physical transformations. So as I take you on the journey through the different shortcuts mathematicians have discovered, each chapter includes a pit stop to explore the shortcuts, or lack of them, in different fields of human activity.

Gauss's schoolroom success at adding the numbers from 1 to 100 using his cunning shortcut fueled his desire to pursue his mathematical talents. His teacher, Herr Büttner, wasn't up to the task of cultivating the budding young mathematician, but he had an assistant, seventeen-year-old Martin Bartels, who was equally passionate about mathematics. Although Bartels had been employed to cut quill pens for the students and assist them in their first attempts at writing, he was more than happy to share his mathematical texts with the young Gauss. To-gether they explored the mathematical terrain, enjoying the shortcuts that algebra and analysis provided to reach their destinations.

Bartels soon realized that Gauss needed a more challenging environment to test his skills. He managed to get Gauss an audience with the Duke of Brunswick. The Duke was so taken by the young Gauss that he agreed to become his patron, funding his education at the local college and then at the University of Göttingen. It was here that Gauss began to learn some of the great shortcuts that mathematicians had developed over the centuries and which would soon become the springboard for his own exciting contributions to mathematics.

This book is my curated guide through two thousand years of better thinking. It has taken me decades to learn how to navigate these cunning tunnels or hidden passes through the landscape, and it took mathematicians through history thousands of years to piece them together. But in this book I've tried to distill some of these clever strategies for attacking the complex problems we encounter in everyday life. This is your shortcut to the art of the shortcut.

CHAPTER 1

THE PATTERN SHORTCUT

> Puzzle: You have a flight of stairs in your house with
> 10 steps. You can take one or two steps at a time.
> For example, you could do 10 one-steps to get to
> the top, or 5 two-steps, or combinations of one-
> steps or two-steps. How many different possible
> combinations are there to get to the top?
>
> You could do this the long way and try to find all
> the combinations, running up and down the stairs.
> But how would our young Gauss do it?

WANT TO KNOW A shortcut to getting an extra 15 percent salary
for doing exactly the same work? Or perhaps a shortcut to
growing a small investment into a large nest egg? How about a shortcut
to understanding where a stock price might be heading in the coming
months? Do you feel like you are sometimes reinventing the wheel
again and again, yet sense there is something that connects all these
different wheels you are making? What about a shortcut to help you
with your terrible memory?

I'm going to dive in and share with you one of the most potent
shortcuts that humans have discovered. It is the power of spotting a
pattern. The ability of the human mind to glean a pattern in the chaos
around us has provided our species with the most amazing shortcut:
knowing the future before it becomes the present. If you can spot a

17

pattern in data describing the past and the present, then by extending that pattern further you have the chance to know the future. No need to wait. The power of the pattern is for me the heart of mathematics and its most effective shortcut.

Patterns allow us to see that even though the numbers might be different, the rule for how they grow can be the same. Spotting the rule underlying the pattern means that I don't have to do the same work every time I encounter a new set of data. The pattern does the work for me.

Economics is full of data with patterns that, if read properly, can guide us to a prosperous future—although, as I shall explain, some patterns can be misleading, as the world witnessed with the financial crash of 2008. Patterns in the number of those falling ill with a virus mean we can understand the trajectory of a pandemic and intervene before it kills too many people. Patterns in the cosmos allow us to understand our past and our future. Looking at the numbers that describe the way stars are moving away from us has revealed a pattern that tells us our universe began in a big bang and will end with a cold future called heat death.

It was this ability to sniff out the pattern in astronomical data that launched the aspiring young Gauss onto the world stage as the master of the shortcut.

Planetary Patterns

On New Year's Day, 1801, an eighth planet was detected orbiting around the sun somewhere between Mars and Jupiter. Christened Ceres, its discovery was regarded by everyone as a great omen for the future of science at the beginning of the nineteenth century.

But excitement turned to despair a few weeks later, when the small planet (which was in fact just a tiny asteroid) disappeared from view near the sun, lost among a plethora of stars. The astronomers had no idea where it had gone.

Then news arrived that a twenty-four-year-old from Brunswick had announced that he knew where to find this missing planet. He told the astronomers where to point their telescopes. And lo, as if by magic, there was Ceres. The young man was none other than my hero Carl Friedrich Gauss.

Since his classroom successes at age nine, Gauss had gone on to make numerous fascinating mathematical breakthroughs, including the discovery of a way to construct a 17-sided figure using only a straightedge and compass. This challenge had been outstanding for two thousand years, ever since the ancient Greeks had started finding clever ways to draw geometric shapes. He was so proud of this feat that he started a mathematical diary, which he filled over the ensuing years with his amazing discoveries about numbers and geometry. But it was the data from this new planet that now fascinated Gauss. Was there a way to find some pattern in the readings that were taken before Ceres disappeared behind the sun that would reveal where to find it now? Eventually he cracked the secret.

Of course, his great act of astronomical prediction was not magic. It was mathematics. The astronomers had discovered the planet by chance. Gauss used mathematical analysis to work out the underlying pattern behind the numbers describing the planet's location, in order to know what it would do next. Gauss was not the first to spot patterns in the dynamic cosmos, of course. Astronomers have been using this shortcut for navigating the changing night sky to make predictions and plan the future ever since our species understood that the future and the past were connected.

Patterns in the seasons meant that farmers could plan when to plant crops. Each season was matched to a particular configuration of stars. Patterns in the behavior of animals migrating and mating allowed early humans to hunt at the most opportune moment, expending the least amount of energy for the most gain. Being able to predict eclipses elevated the status of the predictor within the tribe. Indeed, Christopher Columbus famously exploited his knowledge of an imminent lunar eclipse to save his crew when they were captured by local inhabitants when he became stranded on Jamaica in 1503. So awestruck were the locals by his ability to predict the disappearance of the moon that they acquiesced to his demands for freedom.

What's the Next Number?

For me, the challenge of looking for patterns is perfectly encapsulated in those problems you probably had at school where you are given a sequence of numbers and instructed to determine the next number in the sequence. I used to love the challenges our teacher would chalk up on the blackboard. The longer it took me to spot the pattern, the more rewarding was the experience of uncovering the shortcut. This is a lesson I learned early on. The best shortcuts often take a long time to uncover. It takes work to carve them out. But once revealed, they become part of your repertoire of ways of seeing the world that can be tapped into again and again, like the invention of the printing press or the mechanical loom, tools that can be employed repeatedly to shortcut the production of multiple books or textiles.

To get your pattern shortcut neurons firing, here are a few challenges. What is the next number in this sequence?

$$1, 3, 6, 10, 15, 21 \ldots$$

Not too difficult. You probably spotted that you are just adding on another number each time. So 28 is the next number, because it's 21 + 7. These are called the triangular numbers because they represent the number of stones you need to build a triangle, adding on another row each time. But is there a shortcut to finding the 100th number on this list without having to work your way through all the preceding 99 numbers? This is in fact the challenge that Gauss was faced with when his teacher gave him the task of adding up the numbers from 1 to 100. Gauss found the clever shortcut of adding the numbers up in pairs to get the answer. More generally, if you want the nth triangular number, Gauss's trick translates into the formula

$$1/2 \times n \times (n + 1)$$

These triangular numbers continued to fascinate Gauss after he'd first encountered them in Herr Büttner's class. Indeed, one of the entries of his mathematical diary on July 10, 1796, declares excitedly in Greek, "Eureka!" followed by the formula

$$\text{num} = \Delta + \Delta + \Delta$$

Gauss had discovered the rather extraordinary fact that every number can be written as three triangular numbers added together—for example, $1796 = 10 + 561 + 1225$. This kind of observation can lead to powerful shortcuts because rather than proving that something is true for all numbers, it might be enough to prove it for triangular numbers and then exploit Gauss's discovery that every number is the sum of three triangular numbers.

Here's another challenge. What's the next number in this sequence?

$$1, 2, 4, 8, 16 \ldots$$

Not too tricky: 32 is the next number. This sequence is doubling each time. Called exponential growth, this pattern controls the way a lot of things can grow, and it's important to understand how this kind of pattern evolves. For example, the sequence looks quite innocent to start with. That's certainly what the king of India thought when he agreed to pay the creator of the game of chess the price he demanded for his game. The inventor had asked for a single grain of rice to be placed on the first square of the chessboard and then to double the number of grains of rice on each subsequent square on the board. The first row looked quite innocent, with only a total of $1 + 2 + 4 + 8 + 16 + 32 + 64 + 128 = 255$ grains of rice. Barely enough for a piece of sushi.

But as the king's servants added more and more rice to the board, they very quickly ran out of supplies. To get to the halfway point needs about 280,000 kg of rice. And that's the easy half of the board. How many grains of rice does the king need in total to pay the inventor? At first sight this looks like one of those problems Herr Büttner might give his poor students. There is the hard way to do this: add up 64 different numbers. Who wants to do such hard work? How might Gauss go about this sort of challenge?

There is a beautiful shortcut to making this calculation, but at first sight the shortcut looks like I'm making life harder. Often shortcuts begin by seeming to head in the opposite direction from your destination. First I'm going to give the total grains of rice a name: x. It's one of our favorite names in mathematics, and is in itself a powerful shortcut in the mathematician's arsenal, as I shall explain in Chapter 3.

I am going to kick off by doubling the amount that I am trying to work out:

$$2 \times (1 + 2 + 4 + 8 + 16 + \ldots + 2^{62} + 2^{63})$$

This looks like it's made life more difficult. But stick with me. Let's multiply this out:

$$= 2 + 4 + 8 + 16 + 32 + \ldots + 2^{63} + 2^{64}$$

Now comes the smart bit. I am going to take x away from this. At first sight that looks like I've just got us back to where we started: $2x - x = x$. So how does that help? A bit of magic happens when I replace $2x$ and x by the sums I've got:

$$2x - x = (2 + 4 + 8 + 16 + 32 + \ldots + 2^{63} + 2^{64})$$
$$- (1 + 2 + 4 + 8 + 16 + \ldots + 2^{62} + 2^{63})$$

Most of these terms cancel! There is just the 2^{64} in the first part and 1 in the second part that doesn't get canceled. So all I am left with is

$$x = 2x - x = 2^{64} - 1$$

Instead of lots of calculating, all I need to do is this one calculation to discover that the number of grains of rice that the king needed in total to pay the inventor of chess is

$$18,446,744,073,709,551,615$$

That's more rice than has been produced on our planet in the last millennium. The message here is that sometimes you can play hard work off against hard work and be left with something that is much simpler to analyze.

As the king learned to his cost, doubling starts off looking innocent and then ramps up very quickly. This is the power of exponential growth. The effect is felt by those who take out loans to cover debt. At first sight the offer from a company of a $1,000 loan at 5 percent interest each month might seem like a lifesaver. After one month you only owe $1,050. But the trouble is that each month this gets multiplied by 1.05 again. After two years you already owe $3,225. By the fifth year, the debt is $18,679. Great for the person who's lending money to you, but not so great for the borrower.

The fact that people in general don't understand this pattern of exponential growth means that it can be a shortcut to penury. Payday loan companies have successfully exploited this inability to read the pattern into the future to suck vulnerable people into a contract that initially looks quite attractive. The dangers of doubling and the path it takes us down are important to know before we find ourselves lost and helpless with no way back to safety.

We all learned the frightening rate of growth of the exponential to our cost too late with the pandemic of 2020. The number of people infected doubled every three days on average. And this resulted in healthcare systems being overwhelmed.

On the other hand, the power of the exponential can also help to explain why there are (probably) no vampires on earth. Vampires need to feed on the blood of a human being at least once a month to survive. The trouble is that once you have feasted on the human, the victim too becomes a vampire. So next month there are twice as many vampires in the search for human blood to feast on. The world's population is estimated to be 6.7 billion. Each month the population of vampires doubles. Such is the devastating effect of doubling that within thirty-three months a single vampire would end up transforming the world's population into vampires.

Just in case you ever meet a vampire, here is a useful trick from the mathematician's arsenal to ward off the blood-sucking monster. In addition to the classics—garlic, mirrors, and crosses—one rather unusual way to ward off a vampire is to scatter poppy seeds around his coffin. Vampires, it turns out, suffer from a condition called arithmomania: a compulsive desire to count things. Theoretically, before Dracula finishes trying to count how many poppy seeds are scattered around his resting place, the sun will have driven him back to his coffin.

Arithmomania is a serious medical condition. The inventor Nikola Tesla, whose studies into electricity gave us alternating current, suffered from the syndrome. He was obsessed with numbers divisible by 3: he insisted on 18 clean towels a day and counted his steps to make sure they were divisible by 3. Perhaps the most famous fictional depiction of arithmomania is the Muppets' Count von Count, a vampire who has helped generations of viewers in their first steps along the mathematical path.

Urban Patterns

Here's a slightly more challenging sequence of numbers. Can you sniff out the pattern here?

$$179, 430, 1033, 2478, 5949 \ldots$$

The trick is to divide each number by the number before it. This reveals that the multiplying factor is 2.4. Still exponential growth, but what is intriguing is what these numbers actually represent: patents issued in cities of population size 250,000, 500,000, 1 million,

2 million, 4 million, and so on. It turns out that when you double the population you don't simply get a doubling of the number of patents, as you might expect. Larger cities seem to produce more creativity. The doubling of population seems to add an extra 40 percent to creativity! And it's not just patents that seem to have this pattern of growth.

Despite the huge cultural differences between Rio de Janeiro, London, and Guangzhou, there is a mathematical pattern that connects all cities across the world from China to Brazil. We are used to describing cities by their geography and history, traits that highlight the individuality of a place such as New York or Tokyo. But those facts are mere details, interesting anecdotes that don't explain very much. Look at the city through the eyes of a mathematician, though, and a universal character begins to emerge that transcends political and geographic boundaries. This mathematical perspective unveils the appeal of the city . . . and it proves that bigger is better.

The mathematics reveals that the growth of each resource in a city can be understood by a single magic number particular to that resource. Each time the population of a city doubles, the socioeconomic factors scale not simply by doubling but by doubling and a bit more. Rather remarkably, for many resources that bit more is around 15 percent. For example, if you compare a city with a population of 1 million people to a city of 2 million, then instead of the larger city having twice as many restaurants, concert halls, libraries, and schools, you find an extra 15 percent on top of what you'd expect from simply doubling the numbers.

Even salaries are affected by this scaling. Take two employees doing exactly the same job but in different-sized cities. The employee living in the city with a population of 2 million will on average have a salary 15 percent higher than the salary of the employee in the city with 1 million inhabitants. Double the city size again to 4 million, and the salary gets increased by another factor of 15 percent. The bigger the city, the more you'll get paid for doing exactly the same job.

It's spotting a pattern like this that can be the key to a business getting the most out of what it puts in. Cities come in lots of shapes and sizes. Understanding that the shape is irrelevant but the size matters means that a company can get much more for its buck by simply relocating to a city double the size.

This strange universal scaling was discovered not by an economist or a social scientist but by a theoretical physicist applying the same mathematical analysis that is usually applied in the search for the fundamental laws that underpin the universe. Geoffrey West was born in the United Kingdom, and after studying physics at Cambridge, he went to do research at Stanford exploring properties of fundamental particles. But it was his becoming president of the Santa Fe Institute that would be the catalyst for his discoveries about urban growth. The Santa Fe Institute specializes in finding ways for people from different disciplines to mix and discuss ideas. Very often the shortcut to cracking the enigmas in your own area of research is to take a detour through someone else's seemingly unrelated domain.

It was the mix of mathematics, physics, and biology bubbling away at the institute that led West to wonder whether there were universal characteristics to the cities that span the globe, just as an electron or a photon of light has universal properties regardless of where it is in the universe.

It is possible to believe that mathematics is at the heart of the fundamental laws of the universe, that math can explain gravity or electricity. A city, on the other hand, seems like an incomprehensible mass of people with their own motivations and desires going about the business of their lives. But as we have tried to make sense of the world around us we have discovered that mathematics is the code that controls not only our world and everything in it but even us. There is a pattern to the forces that control even the jumble of millions of individuals.

West and his team amassed data for thousands of cities across the world. They collected everything from the total amount of electrical wire in Frankfurt to the number of college graduates in Boise. They recorded statistics on gas stations, personal income, flu outbreaks, homicides, coffee shops, even the walking speed of pedestrians. Not all the data they needed was on the web, though; West had to grapple with Mandarin in an attempt to decode a massive almanac of data for cities in provincial China.

When they started to analyze the numbers, this hidden code began to emerge. If the population of a city is double that of a smaller city, regardless of where in the world the cities are, social economic factors scale up by this magic number of an extra 15 percent.

Over 50 percent of the world's population now lives in a city. The extra exponential growth provided by this scaling factor could well be the key to why cities are so attractive. You seem to get more out than you put in once a population of people is brought together. This number is probably why people move to big cities—because if a person moves to a city that's twice as big, then all of a sudden they'll get an extra 15 percent more of everything.

Infrastructure too is affected by this scaling, but in the opposite direction. Instead of needing twice as much stuff when you double the size of a city, you find instead that you save on infrastructure. The cost per person of copper wire, pavement, and sewage pipes goes down by 15 percent. Contrary to popular belief, this implies that your personal carbon footprint is smaller the larger the city in which you live.

Unfortunately, it isn't always positive benefits that get scaled by the mathematics. Crime, disease, and traffic get disproportionally scaled by the same factor. For example, suppose you know the number of AIDS cases in a city of 5 million. Then to estimate the number of AIDS cases in a city of 10 million, instead of simply doubling the first number you find you also need to add on an extra 15 percent—the magic 15 percent again.

Is there an explanation for this universal scaling across cities? Is there something like Newton's law of gravity that applies to everything from apples to planets to black holes?

The key to understanding why a city depends on its population size rather than on its physical dimensions is that a city consists not of buildings and roads but of the people who inhabit it. The city is the stage upon which the real actors are playing out the story of civilization. Cities are valuable because they act as a network for facilitating human interactions.

This means we should model a city not by whether it is built on an island, sits in a valley, or sprawls across a desert but by the network of interactions between its inhabitants. It seems to be the quality of the network arising from city interactions that has the property of scaling in this universal manner discovered by West. This is the power of mathematics—to see the simple structures that are at the heart of our complicated environment.

If we take the extreme case that as a city grows everyone has contact with everyone else, then we can see why a large city would lead to superlinear growth. Let's take a measure of the upper limit of connectivity of the inhabitants. If a city has a population of N people, then how many different handshakes can these N people perform? Line them up, numbered from 1 to N. Citizen number 1 shakes all the hands of the people down the line, totaling $N - 1$ handshakes. Now citizen number 2 sets off down the line. They've already shaken citizen 1's hand, so they end up shaking $N - 2$ hands. As we go down the line, each citizen does one less handshake. The total number of handshakes is the sum from 1 to $N - 1$. Hello again! This is the calculation that Gauss was asked to perform. His shortcut produces a formula for this number:

$$1/2 \times (N - 1) \times N$$

What happens to this connectivity when I double N? The number of handshakes doesn't double but goes up by a factor of 2 squared—that is, 4. The number of handshakes is proportional to the square of the number of inhabitants.

This is a great example of why mathematics can spare us from having to continually reinvent the wheel. Although I was asking a completely different question about connections across a network, I found that from the analysis of the triangular numbers I already had the tools to know how this number grows. Time and again the characters might change, but the script remains the same. Understand the script and you've got a shortcut to knowing the behavior of any character inserted into the drama. In this case, the number of connections between citizens grows quadratically with the number of inhabitants.

Of course, there is no way that every inhabitant will know every other citizen in the town. A more conservative measure would be that they know the citizens in their local neighborhood. This would scale linearly; the overall size doesn't really matter.

It looks like real cities are somewhere in between the extreme case and the most limited case. A citizen has all their local connections plus a certain degree of longer-range connections across the city. It seems that the additional long-range connections are the ones that are causing the growth in connections, resulting in the extra 15 percent as the population doubles. As I will explain later in the book, this sort of network arises in many different scenarios and turns out to be a very efficient setup for creating shortcuts across the network.

Misleading Patterns

Although patterns are incredibly powerful, we should still be careful with how we use them. You can set off on a path and think you know where you are heading. But sometimes that path can veer off in a weird and unexpected direction. Take the sequence that I challenged you with earlier in this chapter:

$$1, 2, 4, 8, 16 \ldots$$

What if I told you that 31, not 32, was the next number in this sequence?

If I take a circle, add points on the edge of the circle, and join up all the points with lines, what is the largest number of regions the circle gets divided into? If I have just one point on the circle, then there are no lines and I've got just 1 region. If I add a point, then I can join the two points to get 2 regions, divided by the line I've drawn. Now add a third point. Draw in all the lines connecting points, and I have a triangle figure with three sectors of the circle surrounding the triangle: 4 regions.

Figure 1.1. The first five circle division numbers

If I keep doing this, then it seems like a pattern begins to emerge. Here is the data showing the number of regions as I add another point on the circle:

$$1, 2, 4, 8, 16 \ldots$$

A good guess at this point would be that adding a point doubles the number of regions. The trouble is the pattern breaks down as soon as I add a sixth point on the edge of the circle. No matter how hard you try, the maximum number of regions cut out by the lines is 31. Not 32!

Figure 1.2. The sixth circle division number

There is a formula that will give you the number of regions, but it is a little more complicated than simple doubling. If there are N points on the circle, then the maximum number of regions you'll get when you join the points is

$$1/24 \, (N^4 - 6N^3 + 23N^2 - 18N + 24)$$

The message here is that it is still important to know what your data is describing and not to rely simply on the numbers themselves. Data science is dangerous if it is not combined with a deep understanding of where the data comes from.

Here is another warning about this shortcut. What is the next number in this sequence?

$$2, 8, 16, 24, 32 \ldots$$

Lots of powers of 2 in there. But a slightly unusual 24? Well, if you could identify 47 as the next number in this sequence, then I recommend you buy a lottery ticket next Saturday. These were the winning lottery tickets for the UK National Lottery on September 26, 2007. We are so addicted to looking for patterns that we often see them in places where we can't expect a pattern. Lottery tickets are random. No patterns. No secret formulas. No shortcuts to becoming a millionaire. But that said, I shall explain in Chapter 8 that even random things have patterns in them that we can exploit as potential shortcuts. When it comes to randomness, the shortcut is to stand back and take the long view.

The concept of pattern can be used as a shortcut to understanding when something is truly random or not, and it relates to how memorable a sequence of numbers is.

A Shortcut to a Good Memory

Given that there is so much data being spewed out every second on the internet, companies are on the lookout for clever ways to store it. Finding patterns in the data actually offers a way to compress the information such that you don't need as much space to store it. This is the key behind compression technologies such as JPEG and MP3.

Take a picture that is just black and white pixels. The idea is that in any picture there might be a large swath of white pixels in one corner. Instead of recording each pixel as white and using as much memory to store the picture as there is data in the image, you could take a potential shortcut. Record instead the location of the boundary of the region and just add the instruction to fill in the region with white pixels. The bit of code that I can write to do this will in general be much smaller than recording that each pixel in this region is white.

Any patterns that you can discern in the pixels can be exploited to write code that will record the picture using far less memory than saving the data pixel by pixel. For example, take a chessboard. The image has a very obvious pattern, which allows us to write code that simply says repeat white-black 32 times across the board. Even if you had an enormous chessboard, the code would not grow any bigger.

I believe patterns are also key to how humans store data. I must admit that I have a very bad memory. I think it was one of the reasons I was drawn to mathematics. Mathematics has always been my weapon against my terrible memory for names and dates and random information that I can make no logical sense of. I haven't a clue on what date Queen Elizabeth I died, and if you tell me it was in 1603, I'll have forgotten it ten minutes later; in French I always had difficulty recalling all the different forms of the irregular verb *aller*; in chemistry, was it potassium or sodium that burns purple? But in mathematics I could reconstruct everything from the patterns and logic I'd identified in the subject. Spotting patterns replaced the need for a good memory.

I suspect this is one of the ways our brains store memories. Memory depends on identifying pattern and structure to help our brains store a condensed program from which to regenerate the stored memory. Here's a little challenge. Stare at the squiggles contained in the following 6 × 6 grid. Then close the book. Can you reproduce the grid from memory? The key is not to try to remember each of the 36 squares in the image individually but to find a pattern that helps you to generate the image.

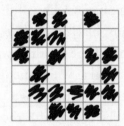

Figure 1.3. Can you memorize where the squiggles are?

Although this image has roughly the same number of squiggles as a 6 × 6 chessboard, the lack of an obvious pattern makes it much harder to remember. The image was generated by tossing a coin and marking a square with a squiggle if the coin landed heads. Mathematically, the coin is equally likely to produce the chessboard pattern of squiggles, alternating between heads and tails, as to produce the random arrangement of squiggles. Yet the pattern in the chessboard makes it much easier to remember.

This idea of the size of the algorithm needed to memorize an image is a powerful measure of the randomness contained inside the image. The chessboard pattern is very ordered. Its generating algorithm is small. The image generated by the toss of a coin probably requires an algorithm that is no shorter than recording the contents of each of the 36 individual squares in the grid.

You will find that a photograph with an obvious story in the picture has a JPEG (a compressed version) that is much smaller than the original, while an image with random pixels fails to get smaller when you try to compress it using the JPEG algorithm because there are no patterns to help it.

Human or machine, anyone or anything that is memorizing something is applying a distinctly mathematical strategy. Memory requires finding patterns, connections, associations, and logic in the data that we are trying to store. Pattern is the shortcut to a good memory.

Climbing the Stairs

Let's return to the question I posed at the start of the chapter. How many ways are there to climb a flight of 10 steps if you use a combination of one-step units and two-step units? There are several ways to go about this. One way would be to just randomly start writing down different possibilities. Clearly an unsystematic approach will inevitably miss some possibilities, and it will also take time to record them all. Is there a better strategy?

A slightly more systematic attack would be to start with just one-steps. There is just one way to achieve this: 1111111111. Next, what if

I allow 1 two-step among the one-steps? There are now a total of 9 steps you will make: 8 one-steps and 1 two-step, and you can choose at which point to make the two-step. There are 9 different places you could do the 2-step.

This is looking like a promising strategy. I could next consider the combinations with 2 two-steps threaded through 6 one-steps, for a total of 8 steps this time. But I am going to have to calculate how many ways there are to choose which of the 8 steps gets a two-step. I could put the first two-step in 8 different locations, and the other two-step in any of the remaining 7 open positions. So that looks like 8×7 different possibilities. But I need to be careful, because I've counted things twice. I could have chosen the first two-step in position 1 and the second in position 2, or I could have done it in the other order. The result would be the same. So the total number now is $8 \times 7 / 2 = 28$. There is actually a mathematical name for this number. It is "8 choose 2," and it is denoted by

$$\binom{8}{2}$$

More generally, the way of choosing 2 numbers from N numbers is given by the formula $1/2 \, (N - 1) \, N$, which is the same formula Gauss came up with for the triangular numbers. There's that wheel we invented showing up again! There is a way to translate the question of choosing 2 numbers from N numbers into the challenge of calculating the triangular numbers. I will explain in Chapter 3 how changing one problem into another can often be a great shortcut to solving a problem.

These tools for calculating choices, called binomial coefficients, were actually some of the formulas that Gauss and his classroom assistant pored over in their algebra books at school together.

But to solve this puzzle, next I will have to calculate how to choose 3 locations out of a choice of 7 to place our 3 two-steps up the staircase.

Although this seems like a good systematic way to build the possibilities up, it is going to require us to generalize these choice functions. It's beginning to look like a hard slog heading along this path. It doesn't really feel like a shortcut.

So here's a better shortcut exploiting what I have shown you in this chapter. With puzzles like this, I find a very powerful strategy is to consider a smaller number of steps and see if there is a pattern in the way the numbers are falling out.

Here are the possibilities for staircases with 1, 2, 3, 4, and 5 steps, which can be worked out quickly by hand:

1 step: 1
2 steps: 11 or 2
3 steps: 111 or 12 or 21
4 steps: 1111 or 112 or 121 or 211 or 22
5 steps: 11111 or 1112 or 1121 or 1211 or 2111 or 122 or 212
   or 221

So the number of possibilities is going 1, 2, 3, 5, 8 . . . Now, you might already have spotted a pattern. You get the next number by adding the two previous numbers together. You might even know a name for these numbers—the Fibonacci numbers, named after the twelfth-century mathematician who discovered that they are the key to the way things grow in the natural world. Petals on flowers, pine cones, shells, populations of rabbits—the numbers all seemed to follow the same pattern.

Fibonacci discovered that Nature was using a simple algorithm in order to grow things. The rule of adding the two previous numbers together was Nature's shortcut for building complex structures like a shell or pine cone or flower. Each organism just uses the two last things it built as ingredients for the next move.

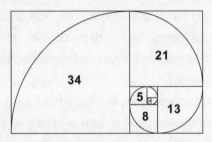

Figure 1.4. How to use the Fibonacci numbers to grow a spiral

Using a pattern to evolve structures is a key shortcut for Nature. Take, for example, the way Nature builds a virus. Viruses come in very symmetrical structures. This is because symmetry requires a simple algorithm to implement to make the structure. If a virus is in the shape of symmetrical dice, the DNA that replicates the molecule just has to make several copies of the same protein that will make up the faces and then the same rule is used across the virus to build its structure. A pattern makes building the virus fast and efficient—and that's part of what can make it so deadly.

But are we really sure just from this small amount of data that the Fibonacci rule is the secret to climbing the stairs?

Actually, the rule explains exactly how to work out the next possibilities with 6 steps on the staircase. Take all the possible steps up to the fourth step and then add a two-step on the end. Or all the possible routes to the fifth step and add a one-step onto these. This gives all the ways up to the sixth step. It is a combination of the two previous numbers in the sequence.

The answer to the puzzle is to calculate the 10th number in the sequence:

$$1, 2, 3, 5, 8, 13, 21, 34, 55, 89$$

There are 89 different routes. The pattern is the shortcut to knowing how many ways to get to the top of the staircase. And the pattern will help crack this conundrum even if there are 100 or 1000 steps instead of just 10.

Although these numbers are named after Fibonacci, he wasn't the first to discover them. In fact, they were first discovered by Indian musicians. Tabla players are interested in showing off the different rhythms they can make on their drum. As they explored the different sorts of rhythms they could make from long and short beats, it led them to the Fibonacci numbers. If the long beat is twice the length of the short beat, then the number of rhythms the tabla player can cook up is the same answer as climbing the staircase. Each one-step corresponds to a short beat, each two-step to a long beat. The number of rhythms the drummer can make of a given duration is the same answer as the number of ways to climb the staircase. So the number of rhythms is given by the Fibonacci rule. And the rule even gives the tabla player an algorithm to construct them out of the previous shorter rhythms.

There is something exciting here about seeing the same pattern explaining so many different things. For Fibonacci, it was the way Nature grows things. For the Indian tabla player, the pattern generates rhythms. The pattern explains the number of ways to climb the staircase in ones and twos. There are some in the financial sector who even believe that these numbers can be used to predict when a stock that is falling will eventually bottom out and start rising again. It is this power of revealing the underlying structure behind different facades that can be so powerful as a shortcut. One pattern solves a multitude of very different-looking challenges. When you are faced with a new problem, it is often worth checking whether it might be an old problem in a new disguise that you already have found a way to solve.

Connecting Shortcuts

I can't resist adding a little coda to this story, because it makes use of the earlier hard work. My first strategy for calculating the number of routes to the top of the staircase started leading me into the question of how to choose 3 things from a group of 7 objects. Mathematicians actually found a clever way to shortcut calculating all those choices I was trying to make. It's something called Pascal's triangle (although, like with the Fibonacci numbers, it turns out that Pascal was beaten to the discovery by the ancient Chinese).

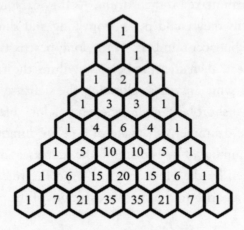

Figure 1.5. Pascal's triangle

The triangle has a rule similar to the Fibonacci rule, but you build the numbers in the layers below by adding the two numbers that sit above that number. The table is easy to build using this rule. But the great fact is that it contains all the choice numbers I was after. Suppose I run a pizza restaurant and I want to boast about the number of different pizzas I offer. If I want to know the number of ways of choosing 3 toppings from a choice of 7 different toppings, then I go to the $(3 + 1)$th number in the $(7 + 1)$th row: 35. That is my shortcut to knowing that there are 35 different pizzas I can make. In general, to choose m things from n things you go to the $(m + 1)$th number in the $(n + 1)$th row. But because these choice numbers were one way to solve our staircase problem, this means that the Fibonacci numbers are actually hiding inside Pascal's triangle. Add up numbers in diagonal lines through the triangle and the Fibonacci numbers appear.

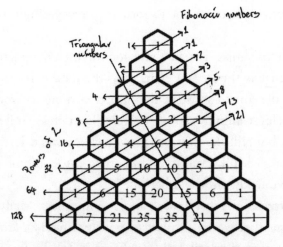

Figure 1.6. The Fibonacci numbers, triangular numbers, and powers of 2 inside Pascal's triangle

This kind of connection is one of the things that I love about mathematics. Who would have thought that the Fibonacci numbers were hiding inside Pascal's triangle? Yet by looking at the puzzle in two different ways I've found a secret tunnel, a shortcut, that links these two seemingly different corners of the mathematical world! And look how the triangular numbers and also the powers of 2 are all hidden inside this triangle. The triangle numbers are sitting along one of the diagonals through the triangle, while you get the powers of 2 by adding up all the numbers in each of the rows. Mathematics is full of these strange tunnels providing shortcuts that we can exploit to change one thing into another.

Finding patterns in data is not just about cute problems of finding ways to climb stairs. It is key to predicting the way the universe will evolve, as Gauss discovered when he predicted the course of Ceres. It is crucial for understanding climate change. It is central to giving businesses an edge as they try to navigate the uncertainties of the future. It might even give us some inkling into the evolution of human history. In this incredibly data-rich age, one exabyte (10^{18} bytes) of data is created on the internet every day. That's a lot of numbers to explore. But spot a pattern and you've got a shortcut to navigating this vast digital landscape.

The pattern shortcut is all about identifying an underlying rule or algorithm that is the key to generating the data you want to understand. It is the sort of shortcut that keeps working for you even if the problem scales seemingly out of control. The staircase might get bigger and bigger, but still the shortcut gets me the answer. It's like boring a tunnel through a hill. Even if the hill grows into a mountain, the tunnel gets me quickly out the other side.

But patterns don't just have to apply to numbers. Many parts of life have patterns that we can exploit to transfer understanding from one realm to another. Understanding patterns in music, for example, is a crucial part of mastering a musical instrument. For Natalie Clein, international cellist, a pattern in music can help her make predictions about where a piece of music might be heading before she's read the notes on the page.

Later in the book I will get the chance to talk to Susie Orbach about shortcuts in psychotherapy. It turns out that she exploits a lot of patterns in human behavior. She can take patterns learned from the history of a previous patient to be able to help a new patient who enters her practice. But people are a little bit messier and more individual than numbers, and the patterns Orbach discerns need to be treated carefully. Where patterns are at their best is when the world is turned into numbers, something our digital world is doing more and more. Our digital footprints are increasingly turning our human behavior into numbers. Find the pattern in the numbers and you have a shortcut to be able to predict what the human might do next.

SHORTCUT TO THE SHORTCUT

Spotting a pattern is an amazing shortcut for navigating your way into the future. Find a pattern in stock prices and you'll have the edge when it comes to investing. Wherever you've got numbers, examine the data for any hidden patterns. But it isn't just numbers that have patterns. People do too. Spot the pattern of shots that your opponent in tennis tends to follow and you'll be ready for the passing shot next time it comes. Understand the pattern in people's eating habits and your restaurant can cater to its customers without excessive wastage of unwanted food. Sniffing out patterns has been a fundamental shortcut for the human species ever since we started taking our first steps out of the savannah.

MUSIC

A FEW YEARS AGO, I DECIDED to learn to play the cello. But it is taking me longer than I had hoped, so I am eager to sniff out any cunning shortcuts that might help. If mathematics is the science of patterns then music is the art of patterns. Could exploiting these patterns be the key?

The cello isn't the first instrument I learned to play. That same year that Mr. Bailson shared the story of the young Gauss with my math class, the music teacher at my school asked the class whether anyone wanted to learn a musical instrument. Three of us put up our hands. At the end of the lesson the teacher led us into the instrument storeroom. It was pretty bare except for three trumpets stacked on top of each other. So the three of us ended up playing the trumpet.

I don't regret the choice. The trumpet is a wonderfully flexible instrument. I cut my teeth playing for the local town band, participated in the local county orchestra, even tried my hand at a bit of jazz. But as I sat counting bars of rest in the orchestra I would stare across at the cellists in front of me, who seemed to be playing all the time. I must admit I was a little envious.

As an adult, I decided that I would buy a cello with a bit of money that my godmother left me in her will. I would use what was left over to take some lessons. But I was slightly concerned whether I'd be up to learning a new instrument. As a child, the time it took to learn an instrument didn't bother me. I was at school and we had years of learning ahead of us. But as adults we have fewer years ahead of us and so become much more impatient. I wanted to be able to play the cello now, not in seven years' time. Was there any shortcut to learning an instrument?

Malcolm Gladwell's book *Outliers* popularized the theory that to become an expert in anything requires putting in a minimum of 10,000 hours of practice. It controversially proposed that this might be enough to become internationally recognized in your field, although the team that produced the original research said that this was a misinterpretation of their work. But was there really no way I could shortcut the 10,000 hours of practice before I could play a Bach cello suite on the stage? An hour a day would mean more than twenty-seven years of practice!

I decided to seek the advice of Naomi Clein, one of my favorite cellists of all time. Clein first came to international attention as one of the youngest winners of the prestigious BBC Young Musician of the Year competition in 1994, when she performed the Elgar cello concerto. What had been her trajectory to international fame?

Clein started playing the cello at the age of six but didn't get serious about it until a few years later. "By fourteen or fifteen," she told me, "I was trying to do four to five hours a day. There are some who do much more. There are kids out there doing about eight hours a day practice when they are sixteen. There are colleagues from places like Russia or the Far East where they're put into this disciplined mode of hard work much earlier than we are in the West."

This level of discipline, Clein explained, was necessary to achieve the motor memory and control that mastering an instrument requires: "There's certainly a minimum number of hours you need to put in when you're learning an instrument, three or four hours a day in your teenage years, which you have to cover, because you physically just don't get the motor mechanics otherwise."

Take Jascha Heifetz, for example. Heifetz was one of the greatest violinists of all time. Famously, he practiced scales every morning for most of his life, thousands of hours in total, just on scales.

In this way, cellists are similar to athletes. You can't run a marathon or win a 100-meter sprint without putting in the hours to physically train your body. The physical aspect of tuning the body and mind so that it can play passages at speed requires brute repetition. I know from my own practice that only after I've repeated a passage over and over, until the body almost knows what to do without the need to engage the brain, that I can play certain pieces.

But Clein was keen to stress that hard work is not simply good enough. "It's repetition of the right thing. It's all very well doing 10,000 hours, but it really has to be the right kind of 10,000 hours. You can't simply do them. What I tell my students is that you have to have mind, body, and soul involved in those 10,000 hours."

Practicing diligently may not seem like a shortcut, but it is. How many times do we waste time doing something because we are doing it the wrong way, not maximizing the effort we are making, or simply missing the point of why we are putting in the hours?

When considering what makes for effective practice, you will often hear people talk about *flow*. The term *flow* was coined by the Hungarian American psychologist Mihaly Csikszentmihalyi in 1990 to describe the psychological state in which we are fully immersed in a task. As he wrote, "The best moments in our lives are not the passive, receptive, relaxing times. . . . The best moments usually occur if a person's body or mind is stretched to its limits in a voluntary effort to accomplish something difficult and worthwhile." Flow lives at the meeting point of extreme skill and great challenge. If you don't have the skills and you try something too challenging, then you end up in a state of anxiety. If something is too easy for you given your skill set, then you are likely to be bored. But if you have both skills and challenge, you can reach this point of flow, or "being in the zone." We'd all love to get to that state, and many people have produced guides to achieving it: meditation, flow soundtracks, dietary supplements, mental flow triggers, caffeine.

But Clein is skeptical of quick fixes. "You can't shortcut your way to the flow. You have to learn the rules in order to be allowed to break them, and it's in the moment of breaking them that you find this liberation that takes you to the flow in some way. It's the discipline that gets you to a state of inspiration."

Although there are no shortcuts to the physical training of being an instrumentalist, I do think that performers spend so much time practicing scales and arpeggios because these provide a shortcut when performing. When you see a pattern of notes on the page that corresponds to a scale or an arpeggio, you don't have to read each note. Instead you can tap into the shortcut of the scales and arpeggios you've spent hours learning.

If there are no shortcuts to achieving the physical fine motor skills required to play at a high level, perhaps there still may be shortcuts to learning a new piece. Clein pointed me to the work of musical analyst Heinrich Schenker. As it happened, I'd come across Schenker before, in a different context. Computer scientists have used his work to try to program AI to compose convincing music. The goal in Schenkerian analysis is to identify a deep structure underlying a piece of music (its *Ursatz*)—a bit like the pattern underlying a sequence of numbers. AI music generation is trying to reverse this process and make music by starting with the *Ursatz* and then fleshing it out. But for Clein, this analysis provides a way to more efficiently navigate a piece of music she is learning.

"[Schenker] likes to reduce and reduce and reduce to the simplest possible formula to understand a piece. You could say that's a shortcut to understanding a structure of a piece of music. It's seeing the macro rather than the micro."

Pattern, it turns out, is part of the musician's tool kit for navigating the complexities of a piece of music. I wondered whether this is a useful shortcut to memorizing a piece of music. For me, identifying the underlying structure of a sequence of numbers allows me to shortcut the need to rely on repetition to memorize something. For Clein, memorizing a concerto just comes from the discipline of practicing something over and over until it becomes motor memory. But for others, patterns may play a bigger role. As Clein told me, "I have a friend, Vadym Kholodenko, who's a kind of genius. I've seen him sight-read a piece in the afternoon that he'd heard once or twice before and then play it in the concert in the evening, and it was really much better than other people had played it after working at it for three months. He sees the big shapes and has an overriding confidence that he can do it and then the rest of the gaps are filled in. He definitely sees the macro and believes in the macro more than the micro, that's for sure."

There is another interesting shortcut that my teacher taught me when I've been learning a new piece. There are often multiple ways to play a passage on the cello because it's possible to play the same note on different strings. Very often the first and most obvious way to play a note is inefficient, and you end up having to jump around the instrument. But if you think more strategically, you can find there are alternative ways to play the same passage that mean you don't have to move your hand wildly back and forth. Working out how to play a piece can become something of a puzzle: What is the most efficient way to place your fingers on the strings so that you can play a passage easily?

Clein agrees: "It can be very creative. I don't think anyone taught me, but I thought it would be a very good idea myself to be able to use my thumb quite a lot. That's been really helpful to me. There are a couple of cellists that do that, starting with the great cellist Daniil Shafran. It was something that I thought I'd made up, but I hadn't. It's all about problem-solving. The more burning the problem is, the more creative the solution can be."

But despite these helpful ways to navigate music, the bottom line for Clein is that there are no shortcuts to what she does. "To become a good professional cellist, especially somebody who has to play solo works and be exposed and feel that you're under the spotlight for your craft, there's just no way. There's no shortcut. And that's what I love about it. Pablo Casals famously practiced all his life, and somebody asked him when he was ninety-five, 'Maestro, why do you keep practicing?' And he replied, 'Because I feel finally I'm getting better. I'm improving.' I think that's what keeps you going. It's a lot of hard work and it continues to be. You have to be interested in the work to keep it going through a lifetime. You don't ever reach a summit."

For many experts, this is why shortcuts aren't a real concern. As Clein told me, "The idea of a shortcut is sort of attractive in the short term, but in the longer term, it's not. I think if there were lots of shortcuts, maybe we'd be less attracted by the challenge."

I recognize that tension between the desire to reach your goal and the ease with which it can realized. If achieving the goal is too easy, it loses its satisfaction. And yet I don't just want mindless slog. The most satisfying shortcuts for me are precisely those that emerge after being stuck for some time, wondering how I'm going to reach my destination. The adrenaline release that accompanies the moment when you see a cunning way through is a drug that I've become quite addicted to in my journey to perfect my mathematical instrument. But when it comes to the cello, although exploiting patterns can help, I realize because of the physical challenges involved there's just no shortcut around hard work.

CHAPTER 2

THE CALCULATED SHORTCUT

> Puzzle: You are a grocer and want to be able to measure all weights from 1 kg to 40 kg using a set of balance scales. What is the smallest number of weights you need to be able to do this, and what are their values?

FINDING THE RIGHT SHORTHAND to capture an idea is a powerful tool for speeding up thought. I take for granted the way that I can capture the concept of a million with seven symbols: 1000000. But bundled up in those seven symbols is a whole history of fascinating shortcuts for how to navigate numbers and how to calculate efficiently. Throughout history, and even today if you are in business or building or banking, you can get an edge if you know some faster, more efficient way to calculate the answer before your competitor does. In this chapter I want to share with you some clever ways that we have found to navigate numbers and calculations. Interestingly, these shortcuts can still be powerful strategies even when there are no numbers involved.

People often think that because I'm a research mathematician, my job must involve doing long division to a lot of decimal places. Surely the calculator should have put me out of a job by now. This misconception of the mathematician as a supercalculator is a common one. But that doesn't mean that calculation is not part of the job description. A lot of inventive mathematics has started from the challenge of finding clever ways to do arithmetic, like Gauss's childhood shortcut. There is a whole history of shortcuts that humans have discovered while trying to more effectively do calculations. Even the calculator we reach for today has been programmed using some of the clever shortcuts mathematicians have come up with over the years.

We tend to think of the computer as all-powerful, able to do anything. But computers have their limits too. Take Gauss's challenge of adding up the numbers to 100. Sure, a computer will have no trouble with that. But at some point there will be a number that is too big even for the computer. If you ask the computer to add up all the numbers to that point, it too will grind to a halt. In general, the computer still relies on us humans to come up with the shortcut that, when implemented in the computer's code, will allow it to get further faster. I shall reveal in this chapter a rather startling use of a seemingly abstruse bit of mathematics, called imaginary numbers, that provided a crucial shortcut for getting computers to do a whole range of tasks, including safely landing planes at an airport.

Shortcut to Counting

The very way that we write our numbers can determine whether a calculation is easy or whether it ends up being a complicated slog prone to errors. It was an important moment for human progress when we realized that good symbols for complex ideas offered a shortcut to better thinking. Historically, it seems that each civilization realized that writing and recording spoken words was a powerful way to preserve, communicate, and manipulate ideas. And every development of a new script for language has generally been accompanied by a clever new way to think about how to record the concept of number. But the civilizations that found better ways to write numbers had a shortcut to faster and more efficient calculations and control of the data.

One of the earliest shortcuts that mathematicians discovered was the power of the place value system. If you are counting something like sheep or days, then the first way you might go about it is to make a mark for each sheep or day. This seems to be how the first humans counted. There are bones dating back forty thousand years with notches down the side that are believed to be examples of humans' first attempt to count.

This is already an impressive moment. The abstract concept of number is beginning to emerge. Archaeologists don't know exactly what these notches are counting, but there is an understanding that the number of notches and the number of sheep or days or whatever was being counted have something in common. The trouble is that trying to distinguish between 17 and 18 notches on a bone is pretty tricky. You have to do the work of counting all over again. At some point in nearly every culture a bright spark hits on the idea of creating a shorthand for all these notches, something that is easier to read.

When I was living in Guatemala a few years ago I was intrigued to find a strange series of dots and dashes on the banknotes. I asked our neighbor whether it was some strange Morse code hidden in the local money. She explained to me it was indeed code, but code for the numbers that each banknote represented. The dots and dashes were shorthand for the way a number was represented in Mayan culture. The Mayans recognized that the human brain has difficulty distinguishing between anything beyond four notches. So instead of recording more and more dots on the page, when they hit five they put a line through the four dots, just like a prisoner counting off the days left to their release. In this way a line becomes shorthand for the number 5.

But what if you want to count even further? The ancient Egyptians came up with an impressive list of hieroglyphs to denote different powers of 10. They represented the number 10 with the image of a hobble for cattle (a device that limits their movement), 100 by a coil of rope, 1,000 by a water lily, 10,000 by a bent finger, 100,000 by a frog, and 1,000,000 by a picture of a man on his knees with his arms in the air looking like he's just won a million in the lottery.

This is a clever shorthand. Rather than representing 1,000,000 by 1,000,000 notches on a bone, the Egyptian scribe could simply draw on the papyrus the figure of the kneeling man. This ability to record large numbers efficiently is one ingredient in the rise of Egypt as a powerful civilization that could tax its citizens and build cities effectively.

But there is still something rather inefficient about the Egyptian system. If a scribe wants to record the number 9,999,999, then they would need sixty-three figures. Add one more to the number and someone has got to come up with yet another little picture to represent 10,000,000. Look how with our modern system of numbers we can record a large number like 9,999,999 with just seven symbols; using just ten different symbols (0, 1, 2, . . . , 9) we can go as far as we want. The key is the *place value system*, an extraordinary shortcut that three different cultures came up with at different points in history.

The first culture to come up with this shortcut was a rival civilization to the Egyptians: the Babylonians. Interestingly, they were not a culture that worked with powers of 10 as the Egyptians did or we do today. Instead they worked with powers of 60. They had numbers all the way up to 59 before they felt the need to regroup. They wrote the numbers from 1 to 59 from just two symbols: $\top$, representing 1, and $\langle$, representing 10. But it meant that the number 59 needed a collection of fourteen symbols.

At first sight this seems far from efficient. But their choice of 60 had a very different sort of shortcut embedded in it. This is a result of the high divisibility of the number 60. Because 60 can be divided in so many different ways—as 2×30 or 3×20 or 4×15 or 5×12 or 6×10—it gave merchants who were using the system many possibilities for how they could divide their merchandise. And the high divisibility of 60 is the reason that we ended up using it for counting time. Our system of 60 minutes in the hour and 60 seconds in the minute has its origins in ancient Babylon.

The Babylonians' real breakthrough, however, was counting beyond 59. One option was to start creating new symbols, the way the Egyptians did. But the Babylonians had a different idea. The meaning of a symbol would change according to its position relative to other symbols. In the system we use, the number 111 has the same symbol repeated three times. But the brilliance of this shorthand is that, reading from right to left, the first 1 represents 1, but the second 1 represents 10, and the third 1 represents 100. Each time you add another number to the left, the value goes up by a factor of 10.

For the Babylonians, however, because they were working in base 60, not base 10, each time you moved to the left the value went up by a multiple of 60. So 111 in Babylonian is $1 \times 60^2 + 1 \times 60 + 1 = 3661$. This was an exceptionally powerful shortcut. Using the two symbols ⊤ and ⟨, you could now represent numbers as big as you wanted. But you couldn't represent every number. This required introducing a new symbol. What if you wanted to record the number 3601? This requires indicating that there are no 60s. In Babylonian cuneiform, the absence of a particular power of 60 was symbolized by two little notches: ⩚.

The Mayans too discovered this shortcut to writing big numbers. They already had the symbol for 5, which was a line. Three lines could denote 15. Three lines and four dots denoted 19. But now the Mayans decided that things were getting too cluttered. And so the next positions in the number denote powers of 20. Thus 111 in Mayan represents $1 \times 20^2 + 20 + 1 = 4041$. They too realized the need for recording the absence of a power of 20 in some places, and the symbol of a shell was used. The Mayans were great astronomers and kept track of huge swaths of time. This efficient number system exploiting the position of a symbol allowed the Mayans to talk about astronomically large numbers without the need for a huge list of symbols.

But there was still something missing in both the Babylonian and Mayan systems: a symbol for nothing. This was the revolutionary step taken by the third culture to invent the place value system: the Indians.

The numbers that we use today are often referred to as Arabic numerals, but this is a misnomer. At the very least it is not the full story—the Arabs learned about the system used by Hindu scribes and brought it to Europe, so the numbers should really be referred to as the

Hindu-Arabic numerals. The Hindu numerals use symbols from 1 to 9, and then each place increases by a factor of 10 as you move left through a number. They also had a symbol for nothing: a zero.

When Europeans were shown this idea, they couldn't get it. Why would you need a symbol if there is nothing to count? But for Indians nothing, the void, was a very important philosophical concept, and so they were happy to name it or number it.

In Europe they were still using Roman numerals and the abacus to do calculations. But using the abacus required skill and expertise. It meant that calculation was not something available to the common citizen. Calculation allowed the establishment to maintain power. There was no record of the calculation done by an abacus, just the result. The system was open to abuse by those in power.

This is why the establishment tried to ban the numerals from the East. They would give the common citizen access to calculation and the ability to record those calculations. The introduction of this shortcut to navigating numbers is probably as significant as the invention of the printing press. It brought math to the masses.

Black Mathematical Magic

Today the computer and calculator are our shortcut to calculation. But those who are over fifty will remember that there was another aid to shortcutting complicated arithmetic: the book of log tables. This was the go-to shortcut for any merchant, navigator, banker, or engineer for centuries. It was the tool that would give them an edge over any competitor who tried to do calculations directly.

The person who unlocked the power of the logarithm was the Scottish mathematician John Napier, who was born in 1550. I would have loved to have met Napier, not just because he came up with this clever calculating shortcut but also because he sounds like a crazy character. Mathematics was not his primary interest; rather, Napier was fascinated by theology and the occult. He would walk around his estate accompanied by a black spider that he kept in a little cage. His neighbors believed he was in league with the devil. When he threatened to imprison a neighbor's pigeons for eating his grain, the neighbor decided to call his bluff, believing it impossible to catch pigeons. The next morning he was shocked to see the pigeons sitting passively in the field and Napier wandering around putting them into sacks. Had they been bewitched? It turns out Napier had drugged them by lacing peas with brandy.

Napier capitalized on the local belief that he was a sorcerer. To catch a thief among his staff he told them that his black rooster could identify the criminal. One by one, the staff had to enter a room and stroke the rooster. Napier claimed the bird would cry out when touched by the thief. When all the staff had visited the rooster Napier asked them to show him their hands. All of them had black soot on their hands except for one. Napier had covered the rooster with soot knowing that only the thief would be too afraid to stroke the bird.

Napier was fascinated by mathematics. But his interest in numbers was nothing more than a hobby, and he lamented the fact that he didn't have sufficient time to carry out his calculations given all his theological studies. But then he came up with a clever strategy to sidestep the long calculations he was trying to wade through.

As he wrote in the book he published about his shortcut:

> Seeing there is nothing (right well-beloved Students of the Mathematics) that is so troublesome to mathematical practice, nor that doth more molest and hinder calculators, than the multiplications, divisions, square and cubical extractions of great numbers, which besides the tedious expense of time are for the most part subject to many slippery errors, I began therefore to consider in my mind by what certain and ready art I might remove those hindrances.

What Napier discovered was a way to turn the complicated job of multiplying two large numbers together into the much simpler task of adding two numbers together. Which of these could you do quicker by hand?

$$379{,}472 \times 565{,}331$$

or

$$5.579179 + 5.752303$$

The key to this magical transformation is the logarithm function. A *function* is like a little mathematical machine that takes one number as input and then manipulates the number according to the internal rules of the function to output a new number. The logarithm function takes a number and outputs the number you would need to raise 10 to the power of in order to get the number you started with. For example, if I input 100, then the logarithm function outputs the number 2, because if I raise 10 to the power 2 I get 100. If I input 1,000,000 into the logarithm function, then the output is 6, because 10 to the power 6 is 1,000,000.

The logarithm function is a bit trickier when it comes to inputting numbers that are not obviously powers of 10. For example, to get the number 379,472 I need to raise 10 to the power 5.579179. To get the number 565,331 I raise 10 to the power 5.752303. So, as with many shortcuts, there was a lot of work that would need to be done in advance to be able to implement the shortcut. Napier spent many hours preparing tables so that he would be able to look up the logarithm of a number, but once the tables were prepared the shortcut came into its own.

If you have two powers of 10—for example, 10^a and 10^b—and you want to multiply them together, then the answer is very simple. It is 10^{a+b}. You just add the powers. This means that instead of doing the hard work of multiplying $379{,}472 \times 565{,}331$ I can add the logarithms $5.579179 + 5.752303 = 11.331482$ and use the tables Napier prepared to calculate the answer: $10^{11.331482}$.

This idea of using tables of calculations to speed up arithmetic wasn't new. Indeed, it seems that some of the cuneiform tablets of the ancient Babylonians were used in a similar fashion. They exploited another formula to work out big multiplications. If I've got two large numbers, A and B, then the algebraic relation

$$A \times B = \tfrac{1}{4} \times \{(A + B)^2 - (A - B)^2\}$$

changed the problem into subtracting square numbers. Instead of calculating the squares, you simply looked them up on one of the tablets of squares that had previously been calculated by a scribe.

Napier described the shortcut he'd come up with in his book *A Description of the Wonderful Table of Logarithms*. And as the book spread it did indeed instill wonder in its readers. Mathematician Henry Briggs, who was the first to hold the prestigious Savilian Professorship in Geometry at my college in Oxford, was so taken by the power of Napier's logarithms that he was eager to meet the man behind the shortcut, and he journeyed four days to visit Napier in Scotland. "I never saw a book which pleased me better or made me more wonder," he wrote.

For centuries these tables would provide scientists and mathematicians with a shortcut to complicated calculations. The great French mathematician and astronomer Pierre-Simon Laplace declared two hundred years later that logarithms, "by shortening the labours, doubled the life of the astronomer and spares him the errors and disgust inseparable from long calculations."

Laplace captures here the essential quality of a good shortcut: it frees up the mind to dedicate its energies to more interesting pursuits. But it was the advent of machines that truly liberated the scientist from the tedium of calculation.

Mechanical Calculators

One of the first to realize the power of machines as a shortcut to calculation was the great seventeenth-century mathematician Gottfried Leibniz. "It is unworthy of excellent men," he wrote, "to lose hours like slaves in the labour of calculation which would safely be relegated to anyone else if machines were used."

Leibniz got his idea for the machine that he would eventually build after encountering a pedometer: "When I saw an instrument with whose help one could count one's steps without thinking, the idea immediately came to me that all of arithmetic could be achieved through a similar type of device."

The pedometer made use of the simple idea that once a cog with ten teeth had clicked through one rotation, it could be connected to another cog that would be clicked one step to record ten steps—it's the place value system in cogs. Called the stepped reckoner, Leibniz's calculating machine was capable of addition, multiplication, and even division. But physically realizing his ideas turned out to be a challenge: "If only a craftsman could execute the instrument as I had thought the model."

He took a wooden prototype to London to give a demonstration to the fellows of the Royal Society. Robert Hooke, who already had a reputation as a cantankerous character, was deeply unimpressed, and after he took the machine apart declared that he could make a much simpler and more efficient contraption. Leibniz was not deterred and eventually managed to employ a skilled clockmaker to make a machine that could achieve the calculated shortcut that he had promised.

Leibniz had an even grander vision. He wanted to mechanize not just arithmetic but all thought. He desired to reduce philosophical argument to a mathematical language that could be implemented on a machine. He envisioned a time when if two philosophers disagreed on some idea, they could simply turn to the machine to sort out their differences and discover who was right.

On a visit to his hometown, Hanover, I was lucky enough to see one of Leibniz's machines. It is a thing of beauty, and we are lucky to have it. For some years, this original machine was stored in the attic of a university building in Göttingen (Gauss's university), and was only rediscovered in 1879 when workers were trying to fix a leak in the roof of the building and came across it hidden in a corner.

Leibniz's machine is the beginning of what would eventually lead to the calculators and computers of today. But that's not to say that there aren't limits to the power of the computer. I suspect we tend to think these days that computers are so good at doing calculations quickly that there are really no limits to what they can do. As *Time* magazine reported in 1984: "Put the right kind of software into a computer and it will do whatever you want it to do." But computers have limits. And even they sometimes need the human programmer to come up with a cunning shortcut to avoid calculations that it would take the computer the lifetime of the universe to execute.

One of the most intriguing shortcuts that computers have tapped into is to exploit a new sort of number that seems to have nothing to do with the practical world of computing: imaginary numbers.

Through the Mathematical Looking Glass

Can you solve the equation $x^2 = 4$? You probably had no trouble coming up with the answer $x = 2$, because if you square 2, you get 4. If you were smart, you might also have come up with a second answer, because $x = -2$ also works. That's because when you square a negative number the answer is positive. So -2 squared is also 4.

That equation was quite simple. But what if I now give you the following equation to solve:

$$x^2 - 5x + 6 = 0$$

This probably sent a shiver down a good number of readers' backs, because this is one of the quadratic equations that students have to learn to solve at school. In fact, the ancient Babylonians had already come up with a general algorithmic procedure that would unlock the answers. Although they didn't yet have the algebraic language to express their ideas, in modern terms if you want to find the solutions to a general equation

$$ax^2 + bx + c = 0$$

then there is a formula that finds the answers:

$$x = \frac{-b \pm \sqrt{b^2 - 4ac}}{2a}$$

So in the case of the equation $x^2 - 5x + 6 = 0$ we put $a = 1$, $b = -5$, and $c = 6$ into the formula and out comes the answer: $x = 2$ or $x = 3$.

It was during the Babylonian period that the power of mathematics to shortcut hard work began to emerge. Before the discovery of this formula, each quadratic equation would be solved by hand. Each time the scribes would be reinventing the wheel, not recognizing that although the numbers were different, what they were doing each time was the same. But at some point a scribe recognized that there was a general algorithmic procedure that worked whatever the numbers.

This is the moment that mathematics begins. It is the art of seeing the pattern that underlies all these infinitely many equations. Instead of a potentially infinite amount of work, the pattern reveals that there is essentially only one act of labor that is needed. Learn the algorithm or formula to solve these equations, and you've got a shortcut to solve infinitely many different equations. With the birth of mathematics in the Babylonian era we witness why mathematics is really the art of the shortcut.

But does this shortcut solve all possible quadratic equations?

How about the challenge of solving the equation $x^2 = -4$? For many centuries this equation was regarded as unsolvable. After all, the numbers that we use for counting have the property that when you square them they are always positive. The Babylonian algorithm or formula wasn't any help because you had to make sense of $\sqrt{-4}$.

But in the middle of the sixteenth century something rather strange happened. In 1551 the Italian mathematician Rafael Bombelli was working on a project to drain the marshes of the Val di Chiana, which belonged to the Papal States. Everything was going well until an interruption suddenly halted the work. With nothing to do, Bombelli decided to write a book on algebra. He'd become interested in exciting new formulas that had emerged to solve equations in a book he'd read by a fellow Italian, Gerolamo Cardano.

The Babylonians had come up with the formula to solve the quadratic equations. But what about cubic equations like $x^3 - 15x - 4 = 0$? Several decades earlier, a number of mathematicians had announced that they had found formulas to solve these cubics. Rather than publishing in scholarly journals, mathematicians at the time enjoyed sparring with each other in public mathematical face-offs. I have a wonderful mental image of heading to the local square on a Saturday afternoon to cheer on your local mathematician in the latest battle of the experts. There was one mathematician whose formula was clearly superior to all the others on the block. The name of this mathematical champion was Niccolò Fontana, better known by his nickname, "Tartaglia." He was understandably loath to give away the secret to his success, but eventually he was persuaded to explain his formula to Cardano on the proviso that Cardano not publish it.

Cardano resisted for several years but then couldn't hold back any longer. The formula appeared in all its glory in his famous book *Ars Magna*, published in 1545. When Bombelli read Cardano's book and applied the formula to the equation $x^3 - 15x - 4 = 0$, something rather odd happened. At some point the formula demanded that he take the square root of -121. Bombelli knew how to take the square root of 121. That was simple; the answer was 11. But what was the square root of -121?

It wasn't the first time that mathematicians had come upon this strange need to take the square root of a negative number, but usually they gave up at this point. Cardano had encountered the same issue and abandoned his calculations. There were no such numbers. But Bombelli kept his nerve. He went on working with the formula that was in Cardano's book, just leaving this strange imaginary number in the formula. And then, as if by magic, the numbers canceled each other out and he was left with the answer that $x = 4$. Sure enough, when he fed that back into the equation, it worked.

In order to reach the final destination of $x = 4$, Bombelli had needed to take a journey through this world of imaginary numbers. It was like stepping through some magic mirror to find a strange new land, and beyond this place, a pathway led to another portal back to the land of normal numbers and the destination you were after. But if you didn't step into this imaginary world, no path existed to a solution. He began to speculate that this wasn't just some trick but that maybe these numbers on the other side of the mirror really did exist after all. It's just that mathematicians needed to have the courage to admit them into their world.

Bombelli's text led to the discovery of imaginary numbers. The most basic number, the square root of -1, was eventually given a name: i. The i stood for *imaginary*, a derogatory term coined some years later by René Descartes, who was far from enamored of these strange and elusive numbers.

And yet Bombelli had revealed their power. In his book he produces a complete analysis of how to manipulate imaginary numbers. If you wanted to solve these cubic equations, then you could take a shortcut to the answer provided you were prepared to step through the looking glass into the world of imaginary numbers. Mathematicians eventually began to refer to these imaginary numbers as *complex numbers*, in contrast to the *real numbers* we all grew up on.

Leibniz was impressed by Bombelli's persistence, declaring him an outstanding master of the analytical art: "Thus we have an engineer, Bombelli, making practical use of complex numbers perhaps because they gave him useful results, while Cardano found the square roots of negative numbers useless. Bombelli is the first to give a treatment of any complex numbers. . . . It is remarkable how thorough he is in his presentation of the laws of calculation of complex numbers."

For centuries, mathematicians remained very suspicious of these numbers. If you wanted the square root of 2, then although it's a number whose decimal expansion is infinite, you still felt you could see this number on a ruler. It was somewhere between 1.4 and 1.5. But where was the square root of -1? You couldn't see that on a ruler. It was my hero Carl Friedrich Gauss who eventually came up with a way to see imaginary numbers.

Before Gauss, the numbers mathematicians had been using were depicted running along a horizontal line, with negative numbers running off to the left, positive numbers to the right. Gauss made the inspired move of heading off in a new direction. These new numbers ran vertically up the page. With Gauss's picture, numbers were no longer 1-dimensional but 2-dimensional. This new map of numbers was extremely powerful. Its geometry reflected the algebraic way these numbers behaved. As I will explain in Chapter 5, a good diagram is an amazing shortcut to explain complex ideas.

Gauss had discovered his picture of these numbers while proving an extraordinary fact about them. If you take any equation, however complicated, made up of powers of x, not just cubes, then these imaginary numbers can always be used to find a solution to the equation. You don't have to make up new numbers. The imaginary numbers were already powerful enough to solve all equations. This great breakthrough of Gauss's is now called the fundamental theorem of algebra.

Gauss's map became a fantastic shortcut to navigate this strange new world of imaginary numbers, but weirdly Gauss kept his 2-dimensional picture a secret. It was later rediscovered independently by two amateur mathematicians, first a Dane called Caspar Wessel and then a Swiss by the name of Jean-Robert Argand. Today the picture is known as the Argand diagram. Credit is rarely fairly attributed.

The French mathematician Paul Painlevé later wrote in his book *Analyse des travaux scientifiques*: "The natural development of this work soon led the geometers in their studies to embrace imaginary as well as real values. It came to appear that, between two truths of the real domain, the easiest and shortest path quite often passes through the complex domain." As well as being a mathematician, Painlevé served as prime minister of France. His first stint in that post, in 1917, lasted only nine weeks but required him to deal with the impact of the Russian Revolution, the entry of the United States into World War I, and quelling a mutiny in the French army.

Even if I'm not explicitly using complex numbers in my work, I often tap into their philosophy. Shortcuts like these are a bit like the wormholes that science fiction writers like to create so that their characters can get from one side of the universe to the other. In any setting, it's worth exploring whether a looking glass might be hidden somewhere that will get you to your goal.

In my mathematical research I'm trying to understand all the symmetries that it is possible to construct. But, strangely, the way I've found to tackle this challenge is to create a new object, called a zeta function, that has its origins in a completely different area of mathematics. And yet it's given me a view on my research that I never would have had if I'd stuck to the world of symmetry. As I will explain in our next pit stop, with entrepreneur Brent Hoberman, the arrival of the internet provided a fantastic looking-glass world in which it's possible to cut out the intermediary in many different commercial transactions.

Sometimes the wormhole that will help me find a way through to a solution can be about simply changing the terrain I am trekking through. When I am stuck on a mathematical problem, I often listen to a piece of music or practice the cello, as a way to let my mind wander. Often when I come back to my desk my view of the problem has altered. The music is like being allowed into the world of imaginary numbers and seeing if, as Painlevé suggests, the path to the desired destination is shorter here. It's worth experimenting with what alternative paths might be out there that can help you access a sneaky trapdoor to a new way of thinking.

Today the world of imaginary numbers is key to understanding a whole range of concepts that would be almost impossible without this shortcut through the looking glass. Quantum physics really only makes sense if encoded in these imaginary numbers. Alternating currents in electronics are most easily manipulated if you describe them using the square root of -1. Another striking example of the shortcut that these numbers provide can be found inside the computers that help land planes in airports across the globe.

BA 107, You Are Cleared for Landing

I was lucky enough some years ago to be given access to the air traffic control tower at one of the United Kingdom's major airports. The screens with mini icons of planes dancing around made it look like a wild computer game. But I was aware that the operators had the lives of many thousands of people in their hands. I was told to be very quiet as I looked on! But when I got a chance to talk to one of the controllers after he'd finished his shift, I was absolutely amazed to find that the system they used to land the planes employed imaginary numbers to speed up the calculations involved in the radar that was tracking the planes as they landed.

It was the German physicist Heinrich Hertz who first discovered that radio waves were reflected by metal objects. He made the discovery during his experiments in 1877 to prove the existence of electromagnetic waves, and he is honored by having his name used for the unit that measures how fast a wave vibrates.

But it was a compatriot of Hertz's who realized the practical possibilities this scientific discovery offered. Christian Hülsmeyer obtained patents in Germany and Britain for an electromagnetic device that he thought could help a ship detect the presence of other ships if visibility was affected by fog. It is said that he was motivated to create such a device after witnessing the grief of a mother who lost her son in a collision of two ships at sea.

He demonstrated his device in an experiment he conducted from a bridge across the Rhine on May 18, 1904. The device would pick up the presence of a boat coming down the Rhine once it was within a 3 km radius. But his equipment was an invention ahead of its time, in part because he hadn't implemented the mathematics to be able to detect how far away the boat was and in which direction. For some years the idea would be one reserved to science fiction by the likes of Jules Verne. The real-world implementation of the idea would take some more decades, and a world war.

Who exactly invented radar (the word originated as an acronym for "radio detection and ranging") is a tricky question. Its development by different countries was kept secret during the buildup to war because it was clear that any country that succeeded in getting it operational would have an edge in detecting incoming planes. But certainly the Scottish physicist Robert Watson-Watt was one of the pioneers of the technology. He'd been asked to comment on rumors of a German death ray based on radio waves. He was quickly dismissive of the idea, but it led him to explore what might be possible with the technology. His demonstration of how the mathematics could be combined with radio signals to track incoming planes led to the establishment of a system of radar stations to detect planes approaching London from the North Sea. His radar network is generally credited with giving the Royal Air Force a crucial edge in the Battle of Britain.

Both in war and in peacetime, speed is of the essence if you are tracking an incoming plane. Finding a shortcut to calculating the location of the aircraft based on the radio waves bouncing back off the planes would be crucial. The basic calculation involved is actually one of trigonometry, a shortcut I will explain in Chapter 4. The shape of the waveforms that are being transmitted and then detected is described by the mathematics of sines and cosines. It turned out that the calculations involved are incredibly tricky and time-consuming. But here is where the discovery of imaginary numbers came to the rescue.

The great eighteenth-century Swiss mathematician Leonhard Euler discovered that if you fed imaginary numbers into the exponential function, the simple function of raising things to the power of a number, then the result was a rather curious-looking combination of wave functions. A century and a half later, it was realized that this output looked very much like the waves that would be used for radar. The connection is actually key to what many mathematicians rank as the most beautiful equation in history, because one instance of this connection between waves and exponential functions produces an equation that links together five of the most important numbers in mathematical history:

$$e^{i\pi} + 1 = 0$$

Instead of having to calculate using the complicated mathematics of wave functions, mathematicians realized you could simplify and speed up the calculations by gluing everything together using imaginary numbers. The use of these strange numbers meant that the calculations were simply ones about exponential functions that could be implemented quickly and efficiently. Even with the extraordinary power of modern computers at their fingertips, air traffic controllers are exploiting this shortcut through imaginary numbers in order to help land planes at airports around the world. Without this shortcut, the planes might have crash-landed before the calculations locating the planes were complete. It's a very graphic illustration of Painlevé's thesis that "between two truths of the real domain, the easiest and shortest path quite often passes through the complex domain."

Binary and Beyond

One of the other shortcuts that computers have exploited on the way to efficient calculating is to use a very economical way of writing numbers. As we have seen, the ten symbols of decimal numbers aren't the only way to represent numbers. We can actually choose powers of any number to represent our numbers, not just 10s as in the decimal system. As we've seen, the Babylonians had symbols all the way from 0 to 59 and worked base 60, while the Mayans had symbols for the numbers from 0 to 19 and created a number system using powers of 20. The choice of 10 for our decimal numbers is purely a quirk of our anatomy, given we have 10 fingers. (I guess since the characters in *The Simpsons* have 8 fingers—except for God, who has 10—that they work in powers of 8.)

It seems that the Babylonian system of base 60 could relate to our anatomy as well. Each finger has 3 knuckles. You can therefore use your thumb on your right hand to point at one of 12 knuckles on the other four fingers of that same hand. Once you have counted a batch of 12 knuckles, you then count this on the left hand, and begin a new count of 12 on the right hand. Given you have 5 fingers on your left hand, you can count 5 lots of 12 knuckles, giving you 60! To represent the number 29 you would raise two fingers on your left hand and use the thumb on your right to point to the fifth knuckle (the middle knuckle of your second finger).

But a computer is limited to just one "finger." Essentially, computers work on the principle of switches being on or off. They need a system that uses just two symbols: 0 for off and 1 for on. Using just these two symbols, the computer can still represent every number. Instead of powers of 10, the places in the place value system represent powers of 2, using what is known as the binary number system. So the number 11011 represents

$$1 \times 2^4 + 1 \times 2^3 + 0 \times 2^2 + 1 \times 2 + 1 = 27$$

Given that we have found ways to translate conversations, pictures, music, and books into digital form, this shortcut using binary has translated the world around us into strings of 0s and 1s.

This idea of binary is also the key to cracking the puzzle that opened this chapter: How few weights can the grocer get away with to measure weights from 1 kg to 40 kg? The trick is to think not in binary but in ternary, or powers of 3. The scales allow three settings: a weight on the right (+1), a weight on the left (−1), or no weight (0). By thinking in ternary, it is possible to show that the grocer only needs four weights that are powers of 3—1 kg, 3 kg, 9 kg, and 27 kg—to measure every possible weight between 1 and 40 kg.

For example, to measure a sack weighing 16 kg, you need to put the sack in one pan along with the 3 kg and 9 kg weights. This will be exactly balanced by putting the 1 kg and 27 kg weights in the other pan. Instead of using 0, 1, and 2 to represent numbers, you are actually using the symbols −1, 0, and 1. So, for example, 16 is represented by

$$1 \ (-1) \ (-1) \ 1$$

which represents 1 unit, minus one lot of 3, minus one lot of 9, plus one lot of 27, making $27 - 9 - 3 + 1 = 16$.

Whether it is numbers or some other complex idea, finding the best notation for representing the concept can be the shortcut to navigating your way through to a solution. The grocer who can think in ternary can get away with purchasing just four weights to do the job. The competitor who hasn't understood this shortcut will find themselves wasting resources on unnecessary weights.

SHORTCUT TO THE SHORTCUT

Finding good shorthand to represent complex concepts has been a crucial shortcut throughout history, not just when it comes to recording numbers. If you are taking notes during a lecture or meeting, then you've probably already started to create shortcuts for key ideas that keep recurring. But might there be a better way to notate your ideas that make them easier to manipulate? Sometimes data in one form is unilluminating, but change the way you are recording it and new insights emerge. Logarithmic graphs often tell us more about data than the original numbers do, which is why, for example, earthquakes are measured using the logarithmic Richter scale. And keep an eye out for a looking glass, like those imaginary numbers, that might take you out of the world you're stuck in and give you an alternative world that provides a shortcut to your destination.

START-UP

USED TO SAY TO MY marketing directors you're really succeeding if you get arrested. None of them managed to do it."

This is what Brent Hoberman, who founded the start-up incubator Founders Factory, told me during a recent visit. Hoberman (who, it must be said, has not yet been arrested) credits pushing the law to its limits for the success of his most famous venture, LastMinute.com, which he co-founded with Martha Lane Fox in 1998. Breaking the rules of the game is part of what Hoberman considers the "entrepreneurial mindset," and that's his shortcut to a successful business venture.

There is a wonderful sense of playfulness about the offices of the Founders Factory. Whiteboards covered in mad scribblings decorate the walls, not so dissimilar to the whiteboards found in math departments all over the world. The open-plan nature of the space means that different start-ups are rubbing shoulders with each other, sharing ideas. Food, drink, and games are available to stimulate ideas. But it is breaking the rules of the game that Hoberman believes is the best shortcut to the success of the ventures being cooked up in the Factory.

"Many entrepreneurs historically broke the rules and then asked for forgiveness later. That was Uber's journey and Airbnb's journey. They're both breaking the law. Why shouldn't people not be able to rent out their own houses? And then society will look at that and say, 'Actually yeah, why not?' That was their shortcut."

Breaking the law is a strategy that has served quite a few mathematicians well. The mathematical laws stated that if you square a number, it must be positive. But Rafael Bombelli had the nerve to start working with a number whose square is −1. By stepping outside the rules of the game, you get access to a whole slew of interesting new mathematics. Euclid states that triangles have angles that add up to 180 degrees. But as we will see later on, mathematicians came up with new geometries where triangles break the Euclidean law. The key to breaking a law is that the benefits are worth the fracture.

As Hoberman explained to me: "It's almost about redefining what this thing is. The regulations may be outdated. It may be that regulations are so slow. Sometimes people are maybe dangerously redefining their own moral compass and saying the trade-off is worth it for society."

The key to the success of LastMinute.com is that it exploited the unused inventory of airlines, car rental companies, and hotels, and then created bundles that are cheaper than buying à la carte. That idea first occurred to Hoberman in his student days when he was looking to treat his girlfriend with a fun weekend away. He would call up hotels at the last minute asking how many suites they had left for the following night. If they said five or six, he knew they were unlikely to sell them all, so he'd offer to take one at a 70 percent discount. "One out of three times it worked."

He began to wonder why everyone wasn't doing this. "They were all too British. The Brits don't do that," he jokes. He remembers banking that experience as a student and realizing that this should be done on an industrial scale. That was the genesis of LastMinute.com. But discovering the unused inventory on an industrial scale required sailing close to the edges of the law. It involved screen-scraping easyJet and Ryanair websites to get their inventory. As Hoberman admits, LastMinute.com technically broke the Computer Misuse Act, which was potentially criminal.

But pushing legal limits is a shortcut that many start-ups have employed to get an edge over competitors. Facebook became famous for its mantra "Move fast and break things." As Mark Zuckerberg once said, "Unless you are breaking stuff, you aren't moving fast enough." Richard

Branson credits his early brush with the law in the 1970s for sparking his business success, although in Branson's case it was because he needed to pay back £60,000 for the tax fraud that he committed in his early days selling records. The fine prompted Branson's much more systematic approach to making money. "Incentives come in all shapes and sizes," Branson wrote, "but avoiding prison was the most persuasive incentive I've ever had."

But as start-ups look toward disrupting highly regulated industries such as healthcare, it becomes more difficult to justify moving fast and breaking things. The healthcare industry works under strict regulatory conditions for obvious reasons. To build up trust in your idea will require working within these conditions. The ethos of "do no harm" outweighs the desire to disrupt. You don't want to be breaking a patient on your way to a successful exit.

One of the other reasons for Hoberman's success was that he took advantage of the amazing shortcut that the internet provided in those early days of the dot-com boom. Time and again it allowed one to cut out intermediaries. In the case of LastMinute.com it was travel agents. Another of his ventures, MADE.COM, exploited a similar shortcut. The idea of MADE.COM was to give consumers access to designer furniture without having to pay designer prices. Hoberman's co-founder, Ning Li, had had his eye on a £3,000 sofa, but then he coincidentally discovered that a friend from school had taken charge of the factory that was making the sofa. They were making it for £250. It sparked the idea of connecting consumers with manufacturers and cutting out the costly intermediary. As Ning says, "There is an elitist mentality in the furniture industry that only customers that can afford £3,000 are entitled to have a trendy, well-made sofa. But there shouldn't be any reason for that." It was the internet that allowed the company to shortcut the supply chain.

Hoberman also recognizes another important shortcut when it comes to setting up companies such as LastMinute.com and MADE .COM: "Ignorance. I would never have started LastMinute.com if I knew how hard it was going to be. You don't want to know too much. Ignorance helps you think differently."

Hoberman's philosophy reminds me of a character in one of my favorite operas. In Wagner's Ring Cycle it is the young Siegfried, who does not know fear, who can successfully kill the dragon Fafnir and take the ring it is guarding. He eventually learns fear when he meets a woman for the first time!

Not knowing fear is, I believe, one of the reasons that the young are perhaps so successful at cracking open big mathematical challenges. Many of us learn a fear of a mathematical beast such as the Riemann hypothesis, our great unsolved problem about prime numbers, that means we think it crazy to try to tackle such a difficult problem: "If generations of mathematicians have failed, then what can I offer?" The dragon remains unslain. What you need is a bit of ignorance mixed with a bit of arrogance. You're not frightened by the history of the problem and you have the self-confidence to think, "Why can't it be me who cracks this great unsolved mystery?"

Hoberman also believes that perfectionism can be a killer for success. If your product is 70 percent ready for launch, then launch it and correct things on the fly. If you wait for it to be 99 percent ready, it will be too late. This philosophy does have its limits. Once other companies started relying on Facebook's platform, for example, it became costlier to just let things crash. Companies might stop using your platform if it is too unreliable. In 2014, Zuckerberg introduced a new philosophy: "Move fast with stable infrastructure." "It may not be quite as catchy as 'Move fast and break things,'" Zuckerberg said with a smirk. "But it's how we operate now."

Perfectionism is considered essential when it comes to mathematics. Most mathematicians believe there is no point even publishing a proof that is only 99 percent complete because that last 1 percent can be deadly. But perhaps we mathematicians are too obsessed with perfectionism. Perhaps it is worth sharing ideas that aren't complete rather than sitting on them and not sharing your thoughts. Newton and to a certain extent Gauss held back progress because they were nervous about sharing incomplete and potentially heretical ideas.

Changing this ethos in the scientific research community is at the heart of Zuckerberg's new initiative, CZI, set up with his wife, Dr. Priscilla Chan. The whole point of CZI is to foster better networks between different research groups, which they believe could produce solutions to medical challenges—solutions that are currently being held back by a fear of sharing research in progress.

Hoberman has gone on to become a big investor in new start-ups, but he still thinks that perfectionism is dangerous when it comes to knowing which companies to back.

"Instinct, I think, is another shortcut. We take shortcuts when we invest in companies. Our best decisions are probably made after a five- or ten-minute meeting. Johannes Reck [started] GetYourGuide, [which] is now a business worth well over a billion. I met him and after ten minutes I said to my colleagues, 'You have to come in and meet him tonight.' Because this guy's got something special. Similarly with Alan.eu, a successful health company in France. I could tell the guy behind it was a genius. I don't need more. Many of my best friends that I tried to bring into that company overanalyzed it."

It is clear from our conversation that Hoberman is a fan of exploiting whatever shortcut will get him to another successful exit.

"I think shortcuts are brilliant, and I chastise my kids when they don't think of shortcuts. Often you'll see people queueing. There are three queues, and everyone just stands in the first one. If you moved to the third one, three meters further away, you'd save ten minutes. But people don't do that. People aren't thinking, 'How do you work out how to get to the front of the queue or find another queue or launch another one?' Life is like a series of those sorts of decisions, and you should always be trying to look for the shortcut."

CHAPTER 3

THE LANGUAGE SHORTCUT

> Puzzle: At Christmas one of the songs I enjoy sing-
> ing is "The Twelve Days of Christmas." "On the first
> day of Christmas my true love sent to me . . . a par-
> tridge in a pear tree." On each subsequent day you
> get the presents of the previous day plus an extra
> number of presents:
>   First day: 1 partridge
>   Second day: 1 partridge + 2 turtledoves
>   Third day: 1 partridge + 2 turtledoves + 3 French
> hens
>   And so on.
>   So by the twelfth day of Christmas, how many
> presents in total did my true love send to me?

ONE OF THE MOST powerful shortcuts I've discovered in my time
as a mathematician is finding the right language to talk about a
problem. Very often the problem will be couched in a language that
obscures what is going on. By finding an alternative language and
translating the problem into a new idiom, the solution suddenly becomes much clearer. A change of language can help you pick out
strange correlations in a company's sales data that the numbers are
obscuring. Much of life is a game, but translating the game into one
you know how to win can give you an amazing edge. And one of the
most exciting revelations I discovered in my apprenticeship as a

mathematician is how a dictionary that changes geometry into numbers provides a shortcut to hyperspace, a multidimensional universe that I have been exploring ever since as a professional mathematician.

There is an increasing number of concepts in science and beyond that don't even seem to exist unless you find the right language to describe them. The idea of emergent phenomena, that qualities emerge from constituent parts, is one such example. Wetness of water is hard to capture if you are talking about individual molecules of H_2O. Although science seems to imply that you can reduce everything to the behavior of fundamental particles and the equations that determine their behavior, this language is often totally inadequate to describe most phenomena. The migration of a flock of birds can't be captured by equations for the movement of atoms that make up the birds. Macroeconomics is rarely comprehensible if you stick to the language of microeconomics. Understanding the effect of an increase in interest rates on inflation is not possible in the language of individual goods even if the microeconomic changes are the cause of the macro phenomenon. Even our ideas of free will and consciousness cannot really be captured by talking about neurons and synapses.

Finding a different language with which to talk about your emotional state can fundamentally change the way you feel. Instead of saying "I am sad," which seems to identify you in a frozen formula that equates you with sadness, you could change this to "Sadness is with me," and suddenly there is the chance for the sadness to move on. As the nineteenth-century American psychologist William James wrote: "The greatest discovery of my generation is that human beings can alter their lives by altering their attitudes of mind." But the power of language doesn't only affect the personal. Language plays a crucial role in the social construction of reality. A society can make things appear by naming them. The concept of a nation-state is as much conjured out of language as it is by geography or a collection of people.

Changing languages sometimes means you can express ideas that are elusive in one language but somehow possible to articulate in another. The fact that German nouns have gender allows one to play games with language that just don't work in English. Heinrich Heine writes of the love of a snowy pine tree for a sunburned Oriental palm. In German a pine tree is masculine and a palm tree feminine, but this nuance is lost in translation to English. Sometimes things get lost the other way. In English you can talk about "his car and her car," but Google Translate will mash this up in French to read "sa voiture et sa voiture" because the car's gender overrides the gender of the owners. Russian has a different word for every possible type of snow- and rainstorm imaginable. Some languages have only five words for colors, while English has many. Pattern, as I've stressed, is an important concept for me, yet when I try to translate the word into French it turns out that there isn't a word that captures the many facets that the word in English embodies.

The significance of differences between languages also fascinated my hero Gauss. At school his teachers had been very impressed by his command of Latin and his lightning mastery of the classics. Indeed, Gauss almost chose to study philology rather than mathematics as he embarked on his studies funded by the Duke of Brunswick. My own journey to become a mathematician followed a not-so-dissimilar journey. I'd wanted to be a spy when I was younger and thought that languages would be an important skill in order to communicate with my fellow agents across the world. At school I'd signed up for all the languages my school taught: French, German, Latin. I even started following a Russian course that was being taught on the BBC. But I was not as successful as Gauss at picking up these new languages, what with their irregular verbs and strange spellings. I became very despondent as my dream for a career in espionage slipped away.

It was when my teacher Mr. Bailson gave me a copy of a book called *The Language of Mathematics* that I began to understand that mathematics was also a language. I think he saw that I was craving a language without irregular verbs, where everything made perfect sense, but he also realized that I wouldn't be able to resist how potent this language was at describing the world around me. In this book I discovered that mathematical equations could tell the story of the planets as they traversed the night sky; that concepts of symmetry could explain the shape of the bubble, beehive, or flower; that number was key to musical harmony. If you wanted to describe the universe, it wasn't German or Russian or English you needed, but mathematics. *The Language of Mathematics* also taught me that mathematics is not just one language but many, and that it is very good at creating dictionaries that translate one language into another, so that shortcuts might appear in the new language. The history of mathematics is punctuated by brilliant moments such as this.

Mathematical Grammar

Implicit in the explanation of many of the patterns that I've shown you so far is an amazing mathematical shortcut: algebra. The trick of algebra is to move from the particular to the general. It means I don't have to keep carving out a new path every time I consider a new case. Instead of considering each particular number in turn, I can let the letter x stand in to mean any number.

Let me do a little trick with you. Think of a number. Double the number. Add 14. Divide the number you've got by 2. Subtract the number you started with. I'll guarantee that the number you're now thinking about is 7. We used this little trick at the beginning of a play that I helped advise on, called *A Disappearing Number*, about the collaboration between Indian mathematician Srinivasa Ramanujan and Cambridge mathematician G. H. Hardy. It always amazed me that this trick elicited a gasp each night from the audience, as if we'd magically read their minds. What has happened, of course, is not magic but mathematics. The key to understand how you've been mathematically manipulated is the idea of algebra.

Algebra is the grammar that underlies the way that numbers work. A bit like a code for running a program, algebra will work whatever numbers you feed into the program. It was developed by the director of the House of Wisdom in Baghdad, a man called Muhammad ibn Musa al-Khwarizmi. Founded in 810, the House of Wisdom was the foremost intellectual center of its time and attracted scholars from around the world to study astronomy, medicine, chemistry, zoology, geography, alchemy, astrology, and mathematics. The Muslim scholars collected and translated many ancient texts, effectively saving them for posterity. Without their intervention, we may never have known about the ancient cultures of Greece, Egypt, Babylonia, and India. However, the scholars at the House of Wisdom weren't content with translating other people's mathematics. They wanted to create a mathematics of their own. It was this desire for new knowledge that led to the creation of the language of algebra.

You can probably spot algebraic patterns on your own, even if you don't know you're doing algebra. As a kid learning my multiplication tables, I started to spot some curious patterns that are hiding underneath these calculations. For example, ask yourself what 5×5 is. Then look at 4×6. What's the connection between the two answers? Now take 6×6 followed by 5×7. Then 7×7 followed by 6×8. Hopefully you've spotted that the second answer is always 1 less than the first.

For me it was spotting these sort of patterns that turned the tedium of rote-learning times tables into something slightly more interesting. But does this pattern always persist? If I take a number and square it, will it always be one more than taking the numbers on either side and multiplying them together?

I've used words to try to describe this pattern, but in the ninth century in Iraq the newly created mathematical language of algebra could describe this pattern. Let x be any number. Then if you square x, it will be 1 more than $(x - 1)$ times $(x + 1)$. Or, written as an algebraic formula,

$$x^2 = (x - 1)(x + 1) + 1$$

This algebraic language also allowed mathematicians to show why this pattern will always persist whatever numbers you choose. Expand $(x - 1)(x + 1)$ and you get $x^2 - x + x - 1 = x^2 - 1$. Add 1 to this and you've just got x^2.

The same approach of letting x stand for any choice of number is the key to the simple magic trick that got you to the number 7. The trick is to translate the instructions into algebra.

Think of a number: x
Double it: $2x$
Add 14: $2x + 14$
Divide by 2: $x + 7$
Subtract the number you first thought of: $x + 7 - x = 7$
And you're now thinking of the number 7.

The point is that it works whatever number you first thought of, even if you were being clever and thinking of an imaginary number! Here's another trick I learned from a mathematical magician friend of mine, Arthur Benjamin. Algebra is key to understanding why the trick works. Throw two dice. Multiply the two numbers you get. Multiply the numbers on the bottom of each die. Then multiply the top of die 1 by the bottom of die 2. Then multiply the bottom of die 1 by the top of die 2. Finally, add up the four numbers you've got. The answer is always 49. What Benjamin has exploited here is the nice fact that the top and bottom numbers on a die always add up to 7. Combine this with some algebra and you always get the answer 49, which is 7 squared.

$$x \times y + (7 - x) \times (7 - y) + x \times (7 - y) + (7 - x) \times y = 7 \times 7 = 49$$

But it wasn't just magic tricks that algebra allowed. The addition of algebra to the mathematician's armory unleashed a huge wave of new discoveries. It was the moment when, instead of just having the words, mathematicians understood the grammar that allowed us to put these words together. It gave us the language to describe how the universe works.

As Leibniz said of the power of algebra: "This method spares the work of the mind and the imagination, in which we must economise above all. It enables us to reason with small cost in effort, by using letters in place of things in order to lighten the load of the imagination."

Lighting the Dark Labyrinth

One person who realized the power of this language to decode Nature was Galileo. He once famously wrote:

> The Universe cannot be understood without first learning to comprehend the language and know the characters in which it is written. It is written in mathematical language, and its characters are triangles, circles and other geometric figures, without which it is impossible to understand a word; without these one is wandering in a dark labyrinth.

One of the stories of the universe that he wanted to be able to read was the challenge of understanding how objects fell to earth. Was there a rule to the way things dropped to the ground or flew through the air? Gathering data from an object dropped from a high building was tricky, as the objects generally fell too fast for anyone to accurately measure their speed. Galileo had a clever idea to slow down the experiment so that he could gather the data he needed. Instead of dropping objects, he would examine the way a ball rolled down a hill. This was slow enough to be able to record how far the ball had rolled after each second.

The incline needed to be smooth enough that the ball wouldn't be slowed by friction. Galileo wanted to approximate as closely as possible a ball falling through space. Once he'd rigged up a smooth surface and started recording the distance the ball would travel each second, he discovered a very simple pattern emerging. If after one second the ball had traveled 1 unit of distance, then in the next second it covered 3 units of distance. Then 5 units of distance in the next second. Each subsequent second the ball was gaining in speed and covering more ground, but the distance it covered was simply following the odd numbers.

It was when Galileo considered the total distance traveled over a period of time that the secret of the way things fall to earth revealed itself.

Total distance traveled after 1 second = 1 unit

Total distance traveled after 2 seconds = 1 + 3 units = 4 units

Total distance traveled after 3 seconds = 1 + 3 + 5 units = 9 units

Total distance traveled after 4 seconds = 1 + 3 + 5 + 7 units = 16 units

Have you spotted the pattern? The total distance is always a square number. But why do odd numbers have anything to do with square numbers? We can find out how by turning numbers into geometry.

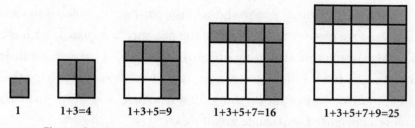

1   1+3=4   1+3+5=9   1+3+5+7=16   1+3+5+7+9=25

Figure 3.1. Connecting square numbers and odd numbers

To grow bigger and bigger squares, I've got to wrap the next odd number in the sequence around the previous square. Suddenly the connection between squares and odd numbers is obvious. Looking at things geometrically rather than arithmetically is a powerful shortcut.

Galileo was now able to cook up a formula to express the total distance that a ball would fall to earth: after t seconds the total distance covered was proportional to the square of t. The fundamental square law of gravity had revealed itself. The discovery of this equation ultimately led to our ability to calculate where a ball would land once it was fired from a cannon and to predict the trajectory of the planets as they orbited the sun.

On the *N*th Day of Christmas

The trick of using a clever geometric way to show the connection between the odd numbers and the square numbers can also be used as a shortcut to solving the puzzle for this chapter. To work out how many presents I get from my true love over Christmas I could go the long route by adding up lots of turtledoves and lords a-leaping. But the shortcut is to change the problem from arithmetic into geometry. Let me start by showing how a geometric perspective can help us to get a hold on the number of presents I get each day. The daily present count is simply the triangular numbers that we encountered in Chapter 1. I've already explained how Gauss cracked these numbers by pairing them up.

But there is another way to shortcut the hard work, which is to view the challenge geometrically. Arrange the presents in a triangle with the partridge at the top. Counting presents making up a triangle is a little tricky. But what if I stick two triangles together? Then I get a shape that is a rectangle. And counting things in a rectangle is easy: it's just base times the height. The triangle is half this.

This geometric shortcut to the solution is essentially Gauss's trick of pairing numbers, just in a slightly different guise. But the geometric perspective allows me to cook up a simple formula to calculate any number in this sequence. If I want the *n*th triangular number, I put two triangles of presents together, which create a rectangle of dimensions $n \times (n + 1)$. Now just divide by 2 to get the number of presents in the triangle: $1/2 \times n \times (n + 1)$.

But what is the total number of presents I get after each day? Here is a running total starting on the first day:

$$1, 4, 10, 20, 35, 56 \ldots$$

You get the next number in this sequence by adding the triangular numbers up in sequence. So to get the next number, the seventh in this sequence, add the seventh triangular number to the previous number. The seventh triangular number is 28, so the seventh number in this sequence is 56 + 28 = 84. But is there a clever shortcut to get to the twelfth number in this sequence, the total number of presents over the whole of the twelve days of Christmas, without adding triangular numbers in sequence?

The trick again is to change numbers into geometry. Think of all the presents coming in boxes of equal size. Then instead of a triangle I can stack all the boxes I've received into a triangular-based pyramid. At the top is a box with a single partridge in a pear tree inside. The next layer down has three boxes: one containing a partridge, the other two containing turtledoves. Each time the day's new presents arrive, I add them to the bottom of the pyramid. So is there any way to understand how many boxes are in the pyramid now that I've changed the number into a shape?

Remarkably, there is. Just as I put two triangles together to make a rectangle, it is possible to put six pyramids of the same size together to make a rectangular stack of boxes. (To make this work you have to do a little shifting around of the way the presents are stacked in each pyramid.) If the pyramid has n layers, then the rectangular structure has dimensions $n \times (n + 1) \times (n + 2)$. But I've got six pyramids making up the structure. The formula for the number of presents in each individual pyramid is

$$1/6 \times n \times (n + 1) \times (n + 2)$$

So how many presents did I get from my true love by the twelfth day of Christmas? Put $n = 12$ into this formula and you get $1/6 \times 12 \times 13 \times 14 = 364$. That's one present for every day of the year except one!

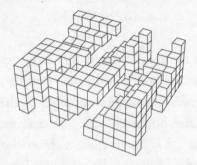

Figure 3.2. Six pyramids make a cuboid

Descartes's Dictionary

I've always loved the way that a picture can make you see something that the numbers obscured. But one needs to be careful. There are times when the eye deceives. Take the following picture.

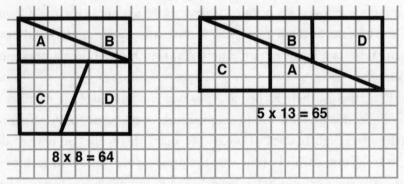

Figure 3.3. Rearrange the pieces to create an extra square

It looks like I've rearranged the pieces that make up the square into a nice rectangle. But hold on. The area of the square is 64, whereas the rectangle has area 65. Where did the extra bit come from? The point is that the picture makes it hard to see that actually the diagonal across the second shape is not truly straight. The edges on the pieces don't line up exactly, and there is just enough room in the gap to accommodate one more unit square. As Descartes famously said: "Sense perceptions are sense deceptions." After I saw this trick, I didn't really trust my eyes ever again. I'm only really happy if I can truly explain a pattern or connection through the language of algebra. What if a similar sneaky trick is going on with those odd numbers I wrapped around the squares?

To reveal these kinds of visual tricks, it is often useful to reverse the shortcut and turn geometry into numbers. Descartes was one of the mathematicians who came up with the idea of a dictionary that translates between numbers and geometry. This dictionary was one of the other great linguistic discoveries, in addition to algebra, that allowed one to find shortcuts to navigating the universe.

We are actually all very accustomed to using this dictionary when using a map or a GPS. The idea is that using a grid layered across a city or country means that I can identify any point in the landscape by two numbers that pinpoint where on this grid the point is located. A GPS uses a grid where the horizontal axis is the equator and the vertical axis is the line of longitude running through Greenwich, England.

For example, if I want to visit the house where Descartes was born, located in a town called Descartes (the town was named after his death, rather than it being an extraordinary coincidence), then the following GPS coordinates will get me there: latitude 46.9726497 and longitude 0.7000201. Every position on the planet can be turned into two numbers.

The power of this translation becomes visible when I want to describe how something might be moving through space. Throw a ball, and I can use two numbers to describe the height above the ground when the ball is a given distance from the person who threw it. The point is that there is a mathematical equation that binds these two numbers together. Suppose x is the distance that the ball has traveled horizontally. Let v be the speed in the vertical direction when I launched the ball and u the horizontal speed. Then if y is the height above the ground, there is a formula for y in terms of these ingredients:

$$y = (v/u)x - (g/2u^2)x^2$$

The letter g stands for a number that is called the gravitational constant. It determines on each planet how hard the ball gets pulled down to the ground by the effect of gravity.

The equation has the same shape however fast or high one launches the ball. One just has to change the values of u and v in the equation. These are like dials that I can twiddle to change the shape of the trajectory. Spotting this pattern underlying the way all balls fly through the air allows us to make predictions about where the ball will land. The equation is a quadratic equation in x. If you are a soccer player and you want to know where to stand in order for an incoming ball to land on your head so that you can knock it into the back of the opponent's net, then you will need to know how to solve this equation for x. As I explained in Chapter 2, two thousand years ago the ancient Babylonians had already found an algorithm for doing this.

But it isn't just the trajectory of balls that these quadratic equations describe. If you look at the price of goods as supply and demand vary, quite often it can be described by the same sort of equations. Knowing how to find the equilibrium point in economics when a good is priced at the point where supply equals demand becomes possible once the numbers are described by the equations. A company that fails to use the language of equations to map the company's data will, as Galileo said, be wandering around in a dark labyrinth while their competitor is raking in the profits.

If you have a set of data points, often you're on the lookout for the equation that might bind them together. Uncover that and you have an amazing shortcut to predicting what might happen next.

It is extraordinary how universal these patterns can be. In the case of throwing a ball, it doesn't matter who throws the ball, how they throw it, where they throw it, or what type of ball they throw—the equation still has the same general shape.

But we have to be careful with fitting equations to data. The population numbers for the United States for the last century can be nicely approximated by a quadratic equation like the one used to trace the trajectory of a ball. But if we use a more complicated equation with terms x^{10}, then we can get an exact fit to the data. That would seem to encourage us to believe the more complex formula is a better predictor. The only trouble is that this equation predicts that the population of the United States plummets to zero in the middle of October 2028. (Or maybe the equation really knows something we don't?)

This story is a word of caution to those who think we can do science just by exploiting the power of big data. Data can suggest patterns, but we still need to combine this with analytic thought to see why the pattern should be dictated by the equation that the data suggests. Galileo's discovery of a quadratic rule behind gravity was subsequently explained thanks to Newton's theoretical analysis, which revealed why quadratics were the right equations to use.

The Shortcut to Hyperspace

The idea of turning geometry into numbers not only allows us to navigate our 3-dimensional universe more efficiently. It also provides a portal to worlds we'll never see. One of the most exciting moments in my mathematical journey through the art of the shortcut was the discovery that I could study hyperdimensional space. The day that I first read about the power of this language to make a cube in 4 dimensions is still etched in my memory.

This explained how a spaceship can get from one end of the universe to the other by taking a shortcut through the fourth dimension. It solves the challenge of how the universe can be finite but without walls. It even allows one to unravel knots that are impossible to untangle in three dimensions.

But it isn't just traveling in space that this dictionary allows. By mapping data into higher-dimensional worlds, hidden structure appears. When you draw a graph of how your data is behaving, you are looking at a 2-dimensional shadow of something that should really be plotted in hyperspace. This shortcut could well reveal subtleties that these 2-D shadows are obscuring. So buckle up as I take you on our journey into hyperspace.

The way to get to the 4th dimension is to start in the 2nd. If I want to describe a square in terms of Descartes's dictionary of coordinates, I can say that it is a shape with four vertices located at the points (0,0), (1,0), (0,1), and (1,1). The flat 2-dimensional world needs just two coordinates to locate each position, but if I also want to include the height above sea level, then I could add a third coordinate. I will need this third coordinate too if I want to describe a 3-dimensional cube in terms of coordinates. Its eight vertices can be described by the coordinates (0,0,0), (1,0,0), (0,1,0), (0,0,1), (1,1,0), (1,0,1), (0,1,1), and (1,1,1).

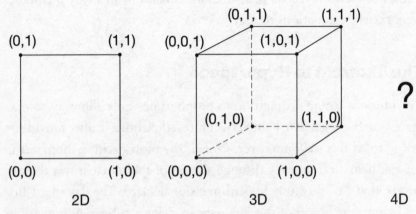

Figure 3.4. Making a hypercube using coordinates

Descartes's dictionary has shapes and geometry on one side and numbers and coordinates on the other. The trouble is that the visual side runs out if I try to go beyond 3-dimensional shapes, since there isn't a 4th physical dimension. It was the great nineteenth-century German mathematician Bernhard Riemann, a student of Gauss's in Göttingen, who recognized that the beauty of Descartes's dictionary is that the other side of the dictionary keeps going.

To describe a 4-dimensional object, I just add a fourth coordinate, which keeps track of how far you are moving in this new direction. Although I can never physically build a 4-dimensional cube, by using numbers I can still describe it precisely. It has sixteen vertices, starting at $(0,0,0,0)$, extending to points at $(1,0,0,0)$ and $(0,1,0,0)$, and stretching out to the furthest point at $(1,1,1,1)$. The numbers are a code to describe the shape. By using this code I can explore the shape without ever having to physically see it.

And it doesn't stop there. You can move into 5, 6, or even higher dimensions and build hypercubes in these worlds. For example, a hypercube in N dimensions will have 2^N vertices. Out of each of these vertices there will be N edges emerging, each of which I am counting twice. So the N-dimensional cube has $N \times 2^{N-1}$ edges.

The taste I got of that 4-dimensional cube whetted my appetite for discovering more shapes in this strange multidimensional universe. It became my passion to carve out new symmetrical objects in this strange world. For example, if you have ever visited the beautiful Alhambra Palace in Granada, then you'll have enjoyed the wonderful symmetrical games that the artists played on the walls of this palace. But is there a way to understand these symmetries? The shortcut for me to understanding at first sight something that appears very visual is to turn symmetry into language.

The creation of a new language for understanding symmetry, called group theory, emerged at the beginning of the nineteenth century. It was the brainchild of an extraordinary young man: the French revolutionary Évariste Galois. His life, alas, was cut tragically short before he could fully realize the potential of his discovery. He was shot dead at age twenty in a duel over love and politics.

Although two walls in the Alhambra may be adorned with very different pictures, the mathematics of symmetry is able to articulate that these two walls have identical symmetries. This is the power of Galois's new language.

Symmetry can be described as the actions that I do to an object that leave it looking like it did before I moved the shape. What Galois understood was that the essential characteristic of symmetry was how the individual symmetries interact with each other—that if you gave the symmetries names, then there was a kind of underlying grammar that interlaced these symmetries. It was this grammar that was the shortcut to unlocking the world of symmetry. The pictures disappeared and in their place was a kind of algebra expressing the way symmetries interact.

With group theory, mathematicians at the end of the nineteenth century were able to prove that there are only seventeen different symmetrical designs that it is possible to draw on the walls in the Alhambra. My own research continues this journey into hyperspace. I am trying to understand the number of ways I might tile an Alhambra in multidimensional space. It's a building made from language, not geometry.

There are ways to get glimpses of these surreal shapes in our mundane 3-dimensional world. La Grande Arche at La Défense in Paris, built by the Danish architect Johann Otto von Spreckelsen, is actually a shadow of a 4-dimensional cube, a cube inside a cube. Salvador Dalí in his painting *Crucifixion (Corpus Hypercubus)* depicts Christ crucified on the 3-dimensional net of a 4-dimensional cube.

There is even a computer game that I've been waiting for many years to be released that promises to give players an experience of living in a 4-dimensional universe. Called Miegakure, the game is the brainchild of designer Marc ten Bosch, who has been making this hypergame for over a decade. The idea is that players faced with a wall that seems to prevent them from advancing in the 3-D environment on the screen can turn toward a 4th dimension and by moving in this new direction find a parallel world that has a shortcut around the side of the wall. The game sounds extraordinary and I can't wait for its release, but I suspect part of the long delay has been the sheer complexity of the 3-D mind of the developer trying to weave together all these 4-D worlds.

Winning at Games

I'm a big fan of games, not just 4-D crazy games. I love collecting games on my travels around the world. But I'm always amazed that even though games from different corners of the globe look very different from each other, often it is the same game just dressed up in a different costume. And this has made me realize that many games become much simpler to play if you can change them into another, apparently different game.

Many of life's challenges are basically games in disguise. A potential cooperation between two rival companies can often turn out to be an example of a game called the prisoner's dilemma. A three-way rivalry can hide a game of rock paper scissors. If you've watched the film *A Beautiful Mind*, you might remember the moment when one of the inventors of game theory, John Nash, played by Russell Crowe, turns the challenge of getting the beautiful woman in a bar into a game. But games have rules that mathematics is very good at navigating. One of the great shortcuts that mathematics has discovered for winning a game is to turn the game into something completely different, where a strategy for winning becomes much more transparent. One of my favorite examples is a game called 15. The players take turns choosing numbers from 1 to 9 with the aim of getting three numbers that add to 15. Once a number is taken the other player can't take it. You have to get to 15 with exactly three numbers—for example, you can pick 1 + 9 + 5, but you can't pick 6 + 9. It's quite a tricky game to play because you have to keep track of the different ways of getting 15 with your numbers and also to stop your opponent from getting to 15 before you. It's worth playing a round with a friend to see how hard it can be to keep abreast of the different possibilities.

But the shortcut to playing this game is to turn it into a completely different game that is easy to play: tic-tac-toe. Except you are going to play the game on a magic square.

2	7	6
9	5	1
4	3	8

The magic square has the property that the sum of the numbers in every row, column, or diagonal is 15. If you play tic-tac-toe on this square, you are in fact playing the game of 15. But the geometry of the game of tic-tac-toe is much simpler to keep track of than the arithmetic of adding numbers up to 15.

Here's another game that becomes easy to play once you spot the right way to look at it. Consider the following map of cities connected by roads. The roads consist of all the straight lines in the map (and so a road can have 2, 3, or 4 cities on it).

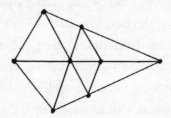

Figure 3.5. Network of roads

You take turns claiming roads. The first person to claim three roads through one city wins. Again it's worth playing to get a feel for the possible strategies. But this is in fact tic-tac-toe again in disguise. If you label the roads with the following numbers, then you are once again playing tic-tac-toe on the magic square.

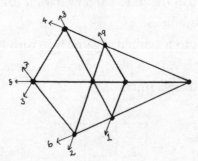

Figure 3.6. Network of roads labeled with a magic square

Another classic game that becomes transparent to play once translated into another language is the game of Nim. There are three piles of beans. Players take turns removing any number of beans from one pile. The winner is the last person to take a bean. You can start with any number of beans in the three piles.

Suppose, for example, that the three piles consist of 4, 5, and 6 beans. Is there a strategy to help you win the game? The trick is to turn the number of beans in each pile into binary. So 4 is 100 and 5 is 101 and 6 is 110. Now there is a strange rule for adding these numbers together that is going to help you see if you are in a winning position or not. Add the numbers in columns, but use the rule that $1 + 1 = 0$. So

$$
\begin{array}{c}
100 \\
101 \\
110 \\
\hline
111
\end{array}
$$

Your strategy is to remove beans from one pile so that this sum changes into 000. It turns out that this is always possible. For example, if I take the pile of 5 beans and remove 3 beans, then there are 2 beans left. In binary, 2 is 010. Do the sum again and we get 000

$$
\begin{array}{c}
100 \\
010 \\
110 \\
\hline
000
\end{array}
$$

But the great thing is that whatever your opponent does next, they have to change this total to something with some 1s appearing. If there are any 1s, they definitely haven't won the game yet. And then you always have a strategy to reset the sum to 000. At some point that will result in you having removed all the beans from the table and winning the game.

The language of binary numbers translates this game into one that you can always win, even if the number of beans or the number of piles changes. If at the outset the sum is already a string of 0s, then make sure you offer to go second; if there are any 1s in the sum, you go first and make a move that sets the sum to 0.

This strategy of using the language of binary numbers to keep track of the state of the game turns out to help us solve a whole slew of similar games. Try playing the following game, called Turning Turtles. There is a line of turtles laid out randomly, with some turtles on their feet and some on their back. (If you don't have turtles at home, then you can use coins. Heads are turtles on their feet, tails are turtles on their back.) Taking turns, players turn a turtle onto its back (or a coin from heads to tails). In addition, if desired, you can turn over one turtle (or coin) to the left of the turtle you put on its back. This turtle or coin can be in either state, on its feet or back (heads or tails). For example, consider the sequence of $N = 13$ coins:

$$T\ H\ T\ T\ H\ T\ T\ T\ H\ H\ T\ H\ T$$

One possible move in this position is to turn the coin in place 9 from heads to tails and then turn the coin in place 4 from tails to heads.

The player to turn the last turtle onto its back or coin from heads to tails wins. At first sight this looks nothing like the game of Nim, but it's actually the same game disguised as turtles.

The number of piles corresponds to the number of turtles still on their feet and the number of items in each pile is the position of the turtle numbered from the left. In the case of the setup with 13 coins, we have 5 piles with 2, 5, 9, 10, and 12 stones in each pile. Turning the turtle on its back (or to tails) in position 9 and also turning the turtle in position 4 on its feet is actually the same as taking 5 stones from the pile with 9 stones. Using the language of binary numbers that wins you Nim now translates into a strategy for turning turtles, a game that at first sight looks totally unrelated.

Although you may never be faced with a game of Turning Turtles, the philosophy at the heart of how to win this game is worth remembering. Faced with any challenge, is there a way to change it into something you already know how to play? Is there a dictionary that you can use that translates the challenge into a language where the solution is more obvious? You might be stuck in a language where there is a wall in the way. But shift into a parallel world by changing the language and a shortcut might just appear that allows you to sneak around the other side of the wall.

SHORTCUT TO THE SHORTCUT

If a problem looks intractable, try to find a dictionary that will translate the description into another language where a solution might reveal itself more easily. If your new DIY enthusiasm isn't being matched by the results you are producing, perhaps you need to change the pictures you are drawing into numbers to see if the measurements reveal why things aren't fitting together as you hoped. If that business plan packed with number-filled tables just isn't communicating its impact, see if a picture or graph helps people grasp your vision. Could a clever bit of algebra actually save you hours of time as you enter your company's finances into yet another spreadsheet? Is your tussle with a competitor actually a game in disguise that you already know a strategy to win? The message of this chapter is to look for the right language that helps you to think better.

MEMORY

ALTHOUGH I'VE SUCCESSFULLY LEARNED THE language of mathematics, it has always frustrated me that I couldn't master those more unpredictable languages like French or Russian that I'd tried to learn in hopes of becoming a spy. Although Gauss too left his love of languages behind to pursue a career in mathematics, he did actually return to the challenge of learning new languages in later life, such as Sanskrit and Russian. At the age of sixty-four, after two years of study, he had mastered Russian well enough to read Pushkin in the original. Inspired by Gauss's example, I've decided to revisit my attempts at learning Russian.

One of the problems I have is simply remembering strange new words. Spotting patterns is my shortcut to memory. But what if there are no patterns? I wanted to know if there might be alternative shortcuts I could try. Who better to ask than Ed Cooke, a Grand Master of Memory and founder of a new venture for learning languages called Memrise?

To earn the title of Grand Master of Memory, one must be able to memorize a 1000-digit number in one hour. The next hour you are faced with memorizing the order of ten packs of cards. And finally you get two minutes to memorize a single pack. In truth, it sounded like a pretty useless skill to try to acquire, but I realized that if I could manage that, then remembering a list of Russian words should be a triviality.

Given that the 1000-digit number is chosen randomly, my strategy of looking for a pattern won't be much help. So what was Cooke's shortcut to 1000 randomly selected digits? It turns out to be something called a memory palace.

"The shortcut is to transform something which is difficult to remember into a proxy that is easier to remember. We remember what is sensory, what is visual, what is tactile, what evokes an emotion. So that's what you want to do, to transform into something like this which recruits your first brainpower," Cooke told me.

"To remember a 1000-digit number, what I do is to arrange a lot of images around a space and every image stands for a number. If I'm trying to recall a number like 7831809720, that's normally a very difficult thing to remember because they're just numbers and they all sound pretty much the same and don't have any meaning. But in my mind 78 is a guy who used to bully me at school and hung me upside down by one leg in my boxer shorts over a staircase, which was a very memorable moment. Much more memorable than the number 78."

Each two-digit number gets turned into a character. In Cooke's private language, 31 is Claudia Schiffer "dressed in memorable yellow underwear from the Citroën advert." This addition of an extra bit of color to the image is important, Cooke says; "the more vivid and bizarre the image, the better the memory." For him, 80 is a friend who has a very amusing face; 97 is the cricketer Andrew Flintoff; 20 is Cooke's father. "I stashed this dictionary of numbers around the age of eighteen, so it's become a fossilized form of my teenage imagination, my humors, beautiful people I'd read about in magazines, my family, my best friends."

Although Cooke is right that for most people one number looks much like another, because I'm a mathematician I know that as one spends more and more time wandering around the world of numbers, one starts to get to know the special characteristics of each number. They start to have their own personalities. It was said of the great Indian mathematician Ramanujan that he knew every number like his own personal friend. When his collaborator Hardy once came to visit him in the hospital and found himself at a loss to offer his ill friend any comforting words, he resorted to recalling that his taxi had had a rather uninteresting number: 1729. Ramanujan immediately replied, "Not so, Hardy, it is a very interesting number. It is the smallest number which can be written as the sum of two cubes in two different ways" (1729 = $12^3 + 1^3 = 9^3 + 10^3$). But most people don't have this kind of intimate emotional relationship with numbers. Claudia Schiffer in yellow underwear is probably more memorable than sums of cubes.

But how does Cooke use this cast of characters to remember a 1000-digit number? The key is to put those characters in space: "If you want to make very, very long chains of information of things, you need a backbone on which to project our images, and it so happens that we have an extraordinary potency of memory for space. Mammals developed an incredible capacity to navigate and remember an incredible repertoire of spaces. Even if we don't think so, we're all really good at this. Just after wandering around an elaborate building for a few minutes we can memorize its layout. So we can use this powerful skill as a shortcut to piggyback our images representing our numbers. This is called building a memory palace."

A memory palace isn't just a story, but a story that moves through space. This last part is key. "The advantage of a memory palace over a simple story is that stories are more vulnerable to breaks in the chain," Cooke explained. "You're also giving yourself the extra burden of making narrative logic rather than just exploiting pure spatial location and therefore it's slightly more taxing on the imaginative mind."

I'd seen Cooke in action building such a palace a few years ago. We'd both taken part in the Serpentine Gallery's Memory Marathon, a weekend exploring the concept of memory, and Cooke took the audience on an amazing walk around the grounds of the gallery using what he saw to create a memory palace that the audience employed to remember all the presidents of the United States. Each name was translated into a powerfully vivid image. For example, President John Adams became the image of Adam and Eve balancing on top of a toilet, *john* being slang for a toilet. These images were then placed at locations around the park. To remember the presidents, the audience just had to mentally resurrect the walk, something that our brains seem very adept at doing, and then use the absurd images located at various points on the walk to recall the presidents.

Using spatial memory seems to be an amazing shortcut to remembering very long sequences, whether it's numbers, presidents, or whatever list you might be trying to commit to memory. It's a fantastic hack, because remembering things by rote seems to scale exponentially in difficulty—the first ten things are easy, the next ten are more difficult, and more than a hundred is almost impossible. But as Cooke explained to me: "The absolutely extraordinary thing about spatial memory is that it seems to scale linearly in difficulty. I can remember a pack of cards in about a minute, perhaps two minutes if I just want to check it over. The thing is that this scales linearly, so in an hour I'd be able to remember thirty packs of cards."

When I suggested that for my readers, perhaps remembering a pack of cards wouldn't really be a skill they were desperate to acquire, Cooke stressed that the cards are irrelevant. The tactic works whatever you are trying to remember. He explained to me that when he is giving talks without notes, he uses exactly the same strategy. Turn the talk into a journey around a familiar place like your house, and place the points you want to make in each of the rooms. Then as you deliver your presentation you will find you can much more easily remember the speech by navigating your way around the memory palace you built in your mind. "In the memory palace, as one embarks on the journey the scene of the action is constantly moving forward, so that the dangers of the memories interfering with each other is thereby diminished because you have a new context in which to excite the new memory."

The technique of translating numbers into visual images is actually key to an extraordinary feat of calculation that my magician friend Arthur Benjamin can do. Benjamin has trained himself to be able to multiply two 6-digit numbers in his head. One of the tricks he used was to exploit a bit of algebra to break 6-digit numbers into pieces that then could be multiplied together separately. But at some point he needs to bank these numbers in memory to be able to continue the calculation, and later these banked numbers need to be recalled for use.

What Benjamin found was that if he simply tried to remember the number, it interfered with his calculations. It's as if number memory was happening in the same place as the calculations. So instead he had a special code that translated numbers into words. Memorizing the words seemed to happen in a part of the brain that wasn't disrupted by further numerical calculations, and so the words could be recalled and translated back into numbers when needed.

My conversation with Cooke took place during the UK lockdown caused by the COVID-19 crisis, and Cooke recalled that it was during another medical lockdown, when as a teenager he found himself in hospital for three months without much to do, that he embarked on his mission to become a Master of Memory. "Partly the motivation was the pleasure of extending a craft to its logical conclusion. As a student my party trick was to remember long numbers and packs of cards in bars in an attempt to win bottles of champagne. And I began boasting to my housemates that I thought I might be one of the world's fastest card memorizers. And they'd retorted, 'Shut up, Ed! Go and prove it.' And that's what took me to these memory championships."

Memory palaces might be good for memorizing strings of digits or delivering speeches without notes, but what about my dream to learn Russian? Is the memory palace technique what Cooke is employing in his company Memrise, his initiative to learn languages? Will I finally find the secret shortcut to learning new languages?

"Repetition and testing," Cooke told me. "By repeating stuff we prove to our brains it's worth remembering. Important things tend to repeat. Testing is crucial because memories are movements of the mind, and those movements become more consolidated the more you practice them."

Those didn't sound like shortcuts, if I'm honest. But Cooke had more.

"The third thing is mnemonics. Say I've got this tricky Russian word, *ostanovka*, which means 'bus stop.' How can I get my head around that word? Well, why don't I relate it to words I know in my own language which will connect it together? If we want to clock something in our mind, we have to weave it into the existing network of associations. So *osta* sounds like Austin, the English car manufacturer. They have made enough cars, giving me *novka*, so I will take the bus, giving me 'bus stop.'"

That sounds more promising. It seems clear that the repetition and testing part means I'm not going to be able to learn Russian in an hour. But mnemonics might actually be a shortcut to remembering a list of Russian words that previously weren't sticking.

Cooke also has one last tip for a shortcut to learning a language, which he learned from his grandmother. "The best way to learn language is between the sheets. If you are enchanted and highly motivated and paying a lot of attention and immersed then you're going to learn extremely fast."

CHAPTER 4

THE GEOMETRIC SHORTCUT

Puzzle: There are ten people in Edinburgh and five in London. The distance between the two cities is 400 miles. Where should they meet so that the total distance traveled by all fifteen people is smallest?

M Y CONCEPT OF THE shortcut for most of this book is an abstract mental shortening of the journey to my destination. But in this chapter I want to consider some real physical shortcuts. If you want to get from A to B in a physical landscape, then understanding the underlying geometry of that landscape can help you to plot paths that get you to your destination faster, even if they look at first sight to be heading in the wrong direction.

Even if you're not actually planning a physical journey, sometimes the challenge you are wrestling with can be translated into something geometric, where a tunnel or a bypass in the geometry translates back to a shortcut in the original problem. For example, digital companies such as Facebook and Google have taken the way large herds of people can collectively find shortcuts through the countryside and translated this philosophy into a way of finding shortcuts in the digital landscape that we stomp through on a daily basis.

Mapping physical shortcuts would also become a passion for Gauss in later years. Although he had fallen in love with mathematics as a student through playing around with numbers, he also enjoyed the challenges of geometry. But this wasn't just the abstract circles and triangles of Euclid. It is rather strange to learn that in his forties, Gauss, a man who loved the abstract ideas of mathematics, signed up for the very practical task of preparing a land survey of the Kingdom of Hanover for the local government. As Gauss once wrote, "All the measurements in the world are not worth one theorem by which the science of eternal truth is genuinely advanced." The work that Gauss engaged in was not the exact and beautiful theory of numbers that he'd been drawn to at school but lots of messy and inaccurate measurements that were full of errors due to faulty equipment or human fallibility. By all accounts, the map of Hanover he ultimately produced was not a particularly accurate one. But the time he spent measuring the state of Hanover did lead to the discovery of revolutionary new sorts of geometry.

Getting from A to B

In 1492 Christopher Columbus famously set sail to find a shortcut to the East Indies. Traditional trade routes required a long and treacherous journey across land that limited the amount of goods traders could bring in any one trip. Traders were eager to find a route by boat. Some believed that there was a route around Africa, although others thought that the Indian sea was landlocked and unreachable this way. Even if there was a way around Africa, many thought that it would take too long. Columbus believed that if he headed west, then he would hit China and India from the other side and hence establish a smoother route for bringing back the spices and silks that Europe wanted from the East.

He'd sat down and done the math. He believed that to get from the Canary Islands to the East Indies would take traveling just 68 degrees west. This, he believed, was a distance of just over 3000 nautical miles—definitely a shortcut if you considered that to sail from London around Africa to the Arabian Gulf was 11,300 nautical miles. Unfortunately, Columbus made several crucial mistakes in his mathematics, which meant that he grossly underestimated the actual distance he would need to travel if he was to go the other way.

Estimates of the circumference of the earth had been made since ancient times. In 240 BCE the Greek mathematician Eratosthenes had calculated it to be approximately 250,000 stadia. How long is a stadion? This is one of the problems with calculating distances—what unit of measurement are you using as your standard? In Eratosthenes's time the unit was a stadion, which was the length of an athletic stadium. The trouble is that Greek stadiums were 185 meters long, while stadiums in Egypt, where Eratosthenes lived and worked, were shorter, at 157.5 meters. Giving Eratosthenes the benefit of the doubt and taking the Egyptian length gives a measurement for the earth's circumference that is just 2 percent off the actual circumference of 40,075 km.

But Columbus took a more recent estimate by the medieval Persian geographer Abu al-Abbas Ahmad ibn Muhammad ibn Kathir al-Farghani, known in the West as Alfraganus. Columbus assumed that the mile that Alfraganus had used in his calculation was the Roman mile, with a length of 4856 feet. Actually, Alfraganus was using an Arabic mile, which was much longer—7091 feet! Fortunately for Columbus, rather than being stuck in the middle of the ocean only halfway to his destination and running out of food and supplies, he stumbled upon a small island in the Bahamas that he named San Salvador. He didn't realize his mistake for some time and referred to the inhabitants of the island as Indians, assuming that he'd reached the East Indies.

The real shortcut to the East eventually turned out to be one that was physically carved out by humans. Already Napoleon during his time in Egypt was flirting with the idea of carving a canal between the Mediterranean and the Red Sea. But due to another set of faulty calculations, it was believed that the Red Sea was actually 10 meters higher than the Mediterranean. To avoid flooding the countries that bordered the Mediterranean, a complex system of locks would need to be built. This ultimately proved too expensive a proposition for the French state.

Once it was realized that the levels of the seas were in fact the same, the idea for a canal soon gathered momentum. The shortcut was finally opened on November 17, 1869. Although the canal was under French control, it was a British ship that ended up being the first to pass through. The night before the opening, under cover of darkness and without lights, the captain of the HMS *Newport* sneaked through the waiting flotilla of ships and managed to park the ship at the end of the canal. When everyone awoke to celebrate the opening, they found the *Newport* blocking the exit to the Red Sea. The only way to get ships through was to let the British ship through first. Although officially reprimanded by the Royal Navy, the captain of the *Newport* was privately congratulated by the Admiralty for the publicity coup.

The Suez Canal shortened the distance from London to the Arabian Gulf by 8900 km, reducing the journey by 43 percent. The importance of this shortcut can be judged by the number of times that it has been fought over, most famously when Egyptian president Gamal Abdel Nasser seized the canal from British control in 1956, precipitating the Suez crisis. Today 7.5 percent of shipping around the world passes through the canal, and the traffic earns the Suez Canal Authority, owned by the Egyptian government, $5 billion a year.

An equally important shortcut that spared ships having to navigate around South America's Cape Horn was opened in 1914. The Panama Canal, which connects the Atlantic Ocean to the Pacific Ocean, actually does have several locks that ships must pass through. It's not because the sea levels are different on either side but because it turned out to be too expensive to dig so deep, so ships are raised to an artificial lake that takes them across Panama.

Around the World

Given that the first circumnavigation of the earth didn't occur until the beginning of the sixteenth century, how did Eratosthenes manage to measure the circumference of the planet so accurately in 240 BCE? He obviously couldn't wrap the earth in a measuring tape. What he did instead was to measure a small distance on the earth's surface and then use some clever math to shortcut having to measure the whole thing.

Eratosthenes was the librarian at the great Library of Alexandria and made fascinating contributions to science in a number of fields, from mathematics and astronomy to geography and music. But despite his innovative work, his contemporaries were rather dismissive of his abilities and gave him the nickname "Beta" to indicate he wasn't in the top rank of thinkers.

One of the clever ideas he came up with was a systematic method for producing a list of prime numbers. To find the primes in a list of numbers from 1 to 100, Eratosthenes proposed the following algorithm: take the number 2 and then strike out all the subsequent numbers that are multiples of 2. This can be done simply by moving through the table in steps of 2 striking out the numbers that you encounter. Next, move to the next number after 2 that hasn't been struck off. This is clearly the number 3. Now strike out all the multiples of 3 by stepping through the table in steps of length 3, systematically striking out all the numbers you encounter. Now the method starts to come into its own, because you now move to the next number you haven't struck off the list. The number 4 has been struck off, so the next number is 5. Repeat the method that we've applied to the previous numbers: move through the table in steps of 5, knocking out all the numbers you encounter.

This is the key to the algorithm: move to the next number not struck off previously and then knock out the numbers that are a multiple of this new number by moving through the table striking off numbers in steps corresponding to the new number you've started from. Apply this systematically to the table, and by the time you knock out multiples of 7 you'll have produced a table of primes up to 100.

It's a clever algorithm. It shortcuts having to think very much. It is perfect for a computer to implement. The trouble is that it very quickly becomes a slow method for churning out primes. It's a shortcut to thinking because you can behave like a machine to produce the list. But it isn't the sort of shortcut that I want to celebrate in this book. What I want is a clever strategy to sniff out the primes.

However, I'm going to give Eratosthenes high marks for his calculation for the circumference of the earth because it is inspired. He'd heard about a well in the city of Swenet with the property that on one day each year the sun was directly overhead. On the summer solstice at midday the sun would cast no shadows on the sides of the well, proving that at this moment the sun was directly overhead. Swenet is known today as Aswan and lies not far from the Tropic of Cancer, which is the line of latitude marking 23.4 degrees north, the farthest point north at which the sun can be found directly overhead.

Eratosthenes realized that he could use this information about the location of the sun to do an experiment on this particular day that, in conjunction with some math, would give him the circumference of the earth. He erected a pole in Alexandria, which he believed to be due north of Swenet. (It's actually 2 degrees of longitude off, but it's the spirit of his experiment that I am applauding, not the accuracy.) On the day of the summer solstice, while the sun was directly overhead in Swenet, casting no shadow in the well there, it caused a shadow to be cast by the pole in Alexandria. Measuring the length of the shadow and the length of the stick, he could construct a triangle with lengths in the same ratio and measure the angle in the triangle. This angle would then tell him how far around the circumference of the earth Alexandria was from Swenet. It turned out that the angle was 7.2 degrees, or 1/50 of a complete circle. Now he just needed to know the distance from Alexandria to Swenet.

Rather than walking the distance himself, he employed a professional measurer, called a bematist. The bematist was required to walk in a straight line between the two cities, as any deviation would mess up the calculation. The calculation that was recorded was in terms of the stadion. Alexandria was 5000 stadia north of Swenet. If the distance from Swenet to Alexandria was 1/50 of the complete journey around the globe, that made the circumference of the earth 250,000 stadia. Today we're not sure exactly how many steps this bematist was using to measure his stadia, but as I explained earlier, it was an amazingly good measurement. Using a bit of mathematical geometry, Eratosthenes shortcut the need to hire someone to walk all the way around the planet.

The word *geometry* has its origins in this experiment, because if we break down the name, it is Greek for "measuring the earth": *geo* = "earth" and *metry* = "measuring."

Trigonometry: A Shortcut to the Heavens

The ancient Greeks weren't just using their math to measure the earth. They realized it could be used to measure the heavens too. And the essential tool that made this possible was not a telescope or a sophisticated tape measure but the mathematics of trigonometry.

There is already a hint of this tool at work in Eratosthenes's calculation. Trigonometry is the mathematics of triangles. It is the mathematics that explains the relationship between the angles in the triangles and the lengths of the sides. This mathematics gave the mathematicians of antiquity an extraordinary shortcut to measuring the cosmos without ever leaving the comfort of the surface of the earth.

For example, already in the third century BCE Aristarchus of Samos had used trigonometry to work out the relative distance from the earth to the sun in terms of the distance from the earth to the moon. To accomplish this, all he had to do was to measure the angle from the moon to the earth to the sun—the three points of a triangle—on the day that the moon is half full. This is when the angle from the earth to the moon to the sun is exactly 90 degrees. Then, by constructing a triangle made up of the angle he had measured, he could calculate the ratio of the distance from the earth to the moon compared with the earth to the sun because the ratio was the same for this smaller triangle he'd drawn on the page. The clever realization is that it doesn't matter how big or small the triangle is; the ratio is the same. This ratio is called the cosine of the angle that Aristarchus had measured.

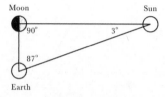

Figure 4.1. Using triangles to measure the solar system

To calculate the actual distance, rather than the ratios of the distances, you need an angle and one length. It was Hipparchus, traditionally credited as the founder of trigonometry, who discovered a cunning way to tease out the actual distances of the earth from the moon and the sun. To do this he exploited a series of solar and lunar eclipses, in particular the solar eclipse of March 14, 190 BCE.

Like Eratosthenes, he used observations at two different locations on the earth. At Hellespont the eclipse was observed as total, but in Alexandria it was a partial eclipse, the moon obscuring just four-fifths of the sun. Combining the distance between the two cities with the angles he measured of the eclipse, he was able to use trigonometry to calculate how far the moon was from the earth.

The power of this trigonometrical shortcut was extraordinary. It led Hipparchus to start preparing the first examples of trigonometrical tables. In these tables you could take an angle, and if you formed a right triangle with this angle at one corner, the table would tell you the ratio of the length adjacent to this angle to the longest length of the triangle. This today is known as the cosine of the angle. Even here mathematicians discovered shortcuts that meant they didn't have to draw out loads of triangles and start doing measurements of lengths and angles.

For example, take an equilateral triangle where all sides are equal and all angles are 60 degrees. Now drop a line from one corner, dividing that angle into 30 degrees; the angle at the base is 90 degrees. The cosine of 60 is easily read off as 1/2, because the adjacent length of this new triangle is half the length of the original side of the equilateral triangle.

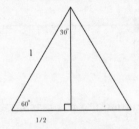

Figure 4.2. The cosine of 60

But then mathematicians discovered a cunning formula that relates cosines in one triangle to cosines in a triangle with half the angle, giving us a tool for additional calculations:

$$\text{Cos}(x)^2 = 1/2 + 1/2\text{Cos}(2x)$$

Using these shortcuts, tables for cosines of many angles could be drawn up. And it was these tables that became the most effective measuring tool for exploring the night sky. The same tables were also key to shortcuts of measurement on earth. It was these that Gauss would have used as he conducted his survey of Hanover. Even surveyors today use this mathematical shortcut to measurement.

For example, if you want to work out the height of a tree, it is quite difficult to lay measuring sticks out from the base of the tree to its top. Instead, a surveyor will walk a distance from the tree and measure the angle from the ground to the top of the tree. Combining this with the much simpler measurement of the distance from the surveyor to the base of the tree and looking up the tangent (which expresses the ratio of the two short sides of the triangle, in this case the height of the tree and the distance from the base to the surveyor), the surveyor gets the height of the tree without climbing any ladders.

A rather beautiful demonstration of the shortcutting power of trigonometry is the measurement of the meter. You might think it a rather curious task to measure a meter, given that this is a unit of measurement. But the story begins with the first definition of what a meter actually meant.

Measuring the Meter

Ever since the first ancient civilizations began building the first cities, we've needed units of measurement to help coordinate construction. The earliest forms of measurement date back to the ancient Egyptians, when body parts were used as units. A cubit was the length from your elbow to the tip of your middle finger. The same use of body parts is evident in pre-metric measurements. A foot is obvious. The word for inch and thumb is the same in many European languages. A yard relates closely to a human pace. Rather intriguingly, the unit of measurement known as a rod, used to measure land in Saxon times, was defined as the total length of the left feet of the first sixteen men to leave church on a Sunday morning. But given that we all come in different shapes and sizes, such measurements will vary from one person to another.

King Henry I tried to resolve this problem by insisting that it was his body that be used to standardize the units. He decreed that the yard should be the distance from the tip of his nose to the end of his out-stretched thumb. But there are clearly problems if you keep on having to compare lengths to a man residing in London.

The leaders of the French Revolution believed that a more egalitarian system of measurement should be put in place, so that everyone could have access to it. Galileo had proved that the swing of a pendulum depends on its length, not its weight or amplitude. It was initially proposed that a meter be the length of a pendulum that takes two seconds to swing back and forth. But it turned out that the swing also depended on the strength of gravity, which varies at different locations around the world.

It was decided instead that a meter be defined as one ten-millionth of the distance from the pole to the equator. Although in principle everyone has access to measuring this distance, the impracticality of using this definition soon became apparent. Two scientists, Pierre Méchain and Jean-Baptiste Delambre, were charged with measuring the distance from the pole to the equator. But just as Eratosthenes realized he could calculate the distance around the circumference of the earth from the distance between Swenet and Alexandria, the two scientists decided to calculate the equator-to-pole distance based on the distance from Dunkirk to Barcelona, two cities roughly on the same line of longitude.

Delambre started out from Dunkirk, in the north, while Méchain was responsible for the southern section, setting off from Barcelona. They agreed to meet at the midway point, Rodez, in southern France. But how did they make their calculations? For a start, they needed a standard length that both of them would use to do the measurements. But even then, they couldn't lay out these lengths along the whole route from Dunkirk to Barcelona.

Here is where the power of trigonometry and triangles kicks in. Delambre, who started at the top of the church tower in Dunkirk, looked across the countryside for two additional high points that could form the other two points of a triangle with the church tower. To start the calculations, he would need to measure the distance from the church tower to one of these points. That hard work couldn't be avoided. But from that point on he could use the measurement of two angles in the triangle to calculate the length of the other two sides of the triangle. To measure angles, he used a piece of equipment called the Borda repeating circle. This consisted of two telescopes mounted on a shared axis with a scale to measure the angle between the two. The way the equipment was used was to point the telescopes at the two high points that Delambre had identified from the top of the church tower and then read off the angle between the telescopes. By moving to one of the other high points in the triangle, he could get the second angle. Then trigonometry takes over to give the lengths of the two missing sides. But here was the really clever thing. One of these sides, which he now knew the length of, would become the side of a new triangle made from picking another high point that he could see from the two high points he'd identified from the church in Dunkirk. In this new triangle he already had one distance. So by measuring two angles with his Borda repeating circle he could calculate the missing distances in the new triangle.

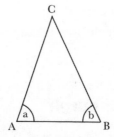

Figure 4.3. Knowing the distance between A and B and
the angles a and b, trigonometry allows you to calculate
the distances from C to A and B.

This was a brilliant shortcut. By rigging up triangles all the way from Dunkirk to Barcelona, the scientists just needed to measure one length in a single triangle and from that point on the angles would do the rest. The science of triangulation is an extraordinary shortcut for surveying the land. Measuring angles can be done from the comfort of the high points marking out the corners of the triangles. No need to pace out the distance or to lay down rulers.

But climbing high points and looking through telescopes wasn't without its dangers. This was not an ideal time to be surveying the land with telescopes and strange equipment. A revolution was raging through the country. The two scientists had to endure many attacks from locals suspicious of these two spying from tops of towers and trees as they measured their way across the country. At Belle Assise, north of Paris, Delambre was arrested on suspicion of being a spy. Why else would he be scaling towers carrying such strange equipment? He tried to explain he was measuring the size of the earth for the Academy of Sciences, but a drunk militiaman interrupted, "There is no more 'Cademy. We're all equal now. You'll come with us." Eventually, after seven years, they returned triumphant to Paris with the meter.

A platinum rod was cast whose length corresponded to their calculation, and from 1799 the standard meter resided in the archives in France. But it suffered the same problem in some sense as Henry I's yard. Despite its universal definition, it was still easier for scientists to journey to France to get a copy of the meter for measurement rather than setting out to make their own measurement of the distance from the pole to the equator.

From London to Edinburgh

When Delambre and Méchain were deciding where to meet up, it was obvious that the halfway point between Dunkirk and Barcelona made sense. But what about the fifteen people in our puzzle that opened the chapter? Where should our team of fifteen people meet up if five are in London, ten are in Edinburgh, and they want to minimize the total distance traveled by all of them? Rather strangely, the answer is that they should all meet in Edinburgh. At first sight you might think that, given the 2-to-1 split in the group, they should meet two-thirds of the way from London to Edinburgh. But for every mile away from Edinburgh, the Scottish group walks an extra 10 miles in total but only saves the English group 5 miles.

More generally, if our fifteen people are scattered at random points along a line from London to Edinburgh, then the shortcut is for them all to head to the position of the person in the middle, the eighth person you encounter heading from London (or Edinburgh). For every mile away from person number 8 you meet, one group saves 7 miles, while the other group has an additional 7 miles to walk (so these even out), but person number 8 adds one more mile to the total.

What about an even more general setting of fifteen people scattered around Manhattan, a city of avenues and streets laid out in a grid? Scanning from east to west, you should meet on the avenue where the eighth person you encounter is standing. But now scan from north to south and choose the street where the eighth person is standing. This will usually be a different person from the eighth when scanned east to west. The meeting spot should be the intersection of the avenue from the east–west scan and the street from the north–south scan.

This kind of analysis is essential if you are trying to find the optimal place for a network exchange for internet cables and you want to minimize the amount of cable you're using. But there is another curious strategy for finding shortcuts through physical and digital spaces that has been exploited throughout history and even in today's technological landscape.

Desire Paths

The explorers of the fifteenth century were after geometric shortcuts that could get them efficiently from one side of the world to the other. In our everyday lives we are often on the lookout for the clever shortcuts that might get us to our destination quicker. In my local park in London the urban planners laid out a crisscross of paved paths to guide the local inhabitants from one side of the park to the other. On paper it probably looked like a perfectly good layout, but the evidence in our park says otherwise. In addition to the paved paths you will find a path worn through the grass where people have decided there is a much quicker route from one side of the park to the other.

Urban planners often like paved paths that make nice right angles to each other, but sticking in a diagonal that cuts out the need to walk around the right angle makes much more sense for pedestrians. Humans prefer the hypotenuse as their path from A to B. Time and again you'll see these paths of flattened grass and bare dirt where people have taken the shortcut to their goal.

An interesting example of these diagonal shortcuts cutting through the right angles can be found in Manhattan. The layout of the streets and avenues running parallel and perpendicular to each other is definitely the mark of human planning. But there is one street that, oddly, cuts diagonally across the grid of roads: Broadway runs from top left to bottom right (northwest to southeast) across the right angles of Manhattan. It turns out that this is actually an ancient shortcut that was used by indigenous travelers before the European settlers appeared on the scene. Broadway follows the Wickquasgeck Path, which is believed to have been the shortest route (avoiding swamps and hills) between Native American settlements that existed at the time. When the

European settlers arrived they kept this shortcut as a route across Manhattan. This path that was trampled down by the feet of travelers going from one side of the island to the other is now preserved in pavement for the cars and pedestrians of the city to use.

These public-made shortcuts have a name: desire paths. Some people call them cow paths or elephant trails because often they would be engraved in the land due to livestock being taken along them. J. M. Barrie, the creator of Peter Pan, describes them as paths that make themselves, because there is never one moment where you see someone laying out the path. No one makes the conscious decision to flatten the grass and clear the way. Such paths emerge gradually, as if making themselves.

Some of these desire paths are rather curious, as they seem to be making the route longer than it should be. They don't look like shortcuts at all. But if you look closely at such a path, you realize that this is a path created to avoid something. Oftentimes it isn't too clear what that is. But dig a bit deeper into the local culture and you'll probably find that some superstition is the key. For example, many people will not walk under a ladder, as it is considered bad luck. They would prefer to go around the ladder. Ladders are often not in place long enough for a permanent desire path to emerge, but in Russia a similar superstition exists for two posts that are leaning against each other. Old streetlights in Russia are often placed at the top of such posts, and you will frequently find that a permanent desire path has emerged to avoid walking between these posts.

Some urban planners have realized that they could use these physical shortcuts as a planning shortcut. Instead of planning the paved paths in advance and finding that people didn't use them, urban planners had the clever idea of letting local inhabitants mark out the desire paths that got them to where they wanted to go, and then, after the paths had emerged, the planners could pave the paths discovered in this organic way.

Michigan State University used the footfalls of students to decide the paths around new university buildings erected in 2011. From an aerial view the paths look like a crazy spaghetti mess of interweaving strands—definitely not something that any designer would have chosen in advance. But having let the students' feet do the talking (or walking), the final layout of paths has created a network that works for all the students trying to get to lectures across the campus. The famous architect Rem Koolhaas used a similar strategy for his design of the campus of the Illinois Institute of Technology in Chicago.

A snowfall also provides an effective way to understand how pedestrians and drivers are using a city. When large swaths of snow lie undisturbed after inhabitants have made their way to the places they want to go, the patterns give the municipality a chance to understand that part of the road or park not being used as a way to traverse the city. This can give urban planners the opportunity to use the land for other purposes, such as a traffic island in a road or a place for a piece of urban art.

This is actually a very common sort of shortcut that you'll find used in the commercial sector over and over: use the public to generate the material from which you can then distill the value. In a sense, the way our digital data is being collected and exploited by companies like Facebook, Amazon, and Google shows that these companies are observing the digital desire paths that we use and then capitalizing on those well-trodden shortcuts.

Twitter, for example, did not introduce the idea of the hashtag in a top-down manner. It was something that the company started seeing employed by users to classify their tweets. In fact, the hashtag seems to have been originated by Chris Messina back in August 2007. He wanted a way to shortcut finding other users interested in the same subjects as he was tweeting about. The hashtag provided a clever way to eavesdrop on conversations of interest. As more and more people followed Messina down this digital desire path, Twitter cottoned on to the shortcut that users had carved out, and it became an official Twitter path—paved over, if you will—in 2009.

Geodesics

If you looked at a map of the world and marked out what you thought was the shortest path to fly from Madagascar to Las Vegas, then your first intuition might be to draw a straight line joining the two across the map. After all, this looks like the desire path that you'd think people (or birds) might fly along. But that doesn't take into account the curvature of the earth. The planet is not flat like a map but a sphere. The true desire path between these two locations, the path of shortest length, actually travels over the United Kingdom and then Greenland.

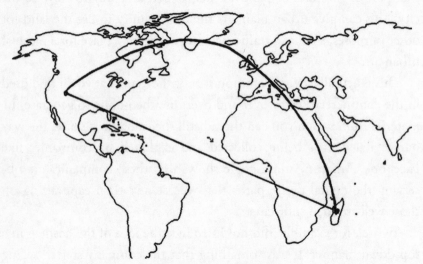

Figure 4.4. The fastest route from Madagascar to Las Vegas
is via the United Kingdom

If you take two points on a sphere, then the shortest path between those two points will be a line called a great circle. This is like a line of longitude that passes through two polar points. In fact, if you took a line of longitude and moved it around the globe until it passed through the two points you were trying to join, then this would be the great circle through these two points.

Once you start to explore the implications of these shortcuts across the globe, some rather interesting features begin to emerge. For example, take three points on the globe—say, the North Pole, Quito in Ecuador, and Nairobi in Kenya. The latter two cities are pretty nearly on the equator. Then the shortest paths between these three points would mark out an example of a triangle on the surface of the earth. Classically, triangles in Euclid's geometry have angles that add up to 180 degrees. But the angles in this triangle on the earth's surface add up to much more than 180 degrees. After all, the two angles at Quito and Nairobi are already each 90 degrees since the line of longitude from the pole meets the equator at 90 degrees. The angle at the North Pole between the two lines of longitude running through these two cities is 115 degrees. So the sum of the angles in the triangle is 90 + 90 + 115 = 295 degrees.

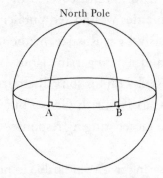

Figure 4.5. On a sphere angles in triangles add
up to more than 180 degrees

There are also geometries where angles in triangles sum to less than 180. For example, in something called an antisphere, which looks like a cone with curved sides, the shortest paths between two points on this surface will trace out strange triangles where the sum of the angles is less than 180 degrees. These surfaces have what is called negative curvature, while spheres like the earth have positive curvature. A flat geometry like the map I started with has curvature zero.

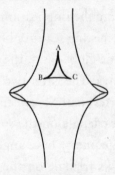

Figure 4.6. On an antisphere angles in triangles add up to
less than 180 degrees

The discovery of curved geometries was one of the exciting developments in mathematics that came in the early nineteenth century. But their discovery caused something of a fight among three mathematicians who each claimed to have discovered the geometries first. The idea of these new geometries was floated publicly for the first time in the 1830s simultaneously by a Russian mathematician, Nikolai Ivanovich Lobachevsky, and a Hungarian, János Bolyai. Bolyai's father was particularly impressed with his son's discovery and boasted of it to a good friend of his: Carl Friedrich Gauss. But Gauss wrote back to Bolyai's father with the rather stinging response:

> If I commenced by saying that I am unable to praise this work, you would certainly be surprised for a moment. But I cannot say otherwise. To praise it would be to praise myself. Indeed the whole contents of the work, the path taken by your son, the results to which he is led, coincide almost entirely with my meditations, which have occupied my mind partly for the last thirty or thirty-five years.

It turns out that indeed Gauss had discovered these curved geometries with strange shortcuts across the surfaces many years earlier, while he'd been making his survey of Hanover. This involved triangulating the land, as Méchain and Delambre had done to measure the meter. Although the task had seemed at first sight a tedious chore for the great mathematician, it turned out to be the catalyst for deep theoretical insights. Gauss had speculated whether not just the surface of the earth but the geometry of space itself might be curved. Gauss decided to use some of his measurement of triangles to test whether light beams traveling between three hilltops around his home in Göttingen might create a triangle whose angles add up to something different from 180 degrees.

Light loves shortcuts. It has the property that it finds the shortest path between two points. So if the angles added up to something different from 180 degrees, it would mean that the light was following a curved path through space. Gauss was hoping to prove that 3-dimensional space was actually curved, just like the 2-dimensional surface of the earth. When he didn't find any discrepancies, he abandoned his ideas, since these new curved geometries went against his belief that mathematics was there to describe the universe we see around us. He swore to secrecy the few friends with whom he did discuss his research.

We now know, of course, that Gauss was working on too small a scale to detect the curvature of space. It was Albert Einstein's new theory of gravity and the geometry of space-time that sparked a renewed interest in testing out Gauss's ideas.

Einstein discovered that the distance between two objects in space could vary according to who was doing the observing. If you were traveling at nearly the speed of light, the distance would look shorter. Time also seemed to depend on the observer. The sequence of events could change according to how the observer was moving. Einstein's big breakthrough was to realize that you needed to consider time and space together in a 4-dimensional geometry made up of three spatial dimensions and one dimension of time. To measure distance in this new space-time geometry led to a shape that was curved.

Einstein's insights redefined gravity not as a force, as Newton had conceived of it, but as a bending of the geometry of space-time. An object with a large mass would warp the fabric of space. Rather than the force of gravity being like a force pulling objects together, one could rethink the story and see gravity as the shortcuts that objects took through this geometry. The free-falling of an object was really just the object finding the shortest path through the geometry from one point to another.

Hence a planet orbiting the sun is not to be thought of as an object pulled by a force, as though a string were attached to it, but rather as a ball rolling down the side of this 4-dimensional space-time geometry. It seemed a crazy idea, but Einstein thought of a way to test it. Light, just like planets, should also find the shortest path through space. If light was due to pass near an object of large mass, the theory implied that the shortest path for the light would be to take a detour, curving toward the object.

The British astronomer Arthur Eddington realized that the solar eclipse that was due to hit earth in 1919 could be used to show whether Einstein was correct. The theory predicted that the light from distant stars should be bent by the gravitational effect of the sun. Eddington needed the eclipse to block out the glare of the sun so that he could actually see the stars in the sky. The fact that light indeed seemed to bend around objects of large mass confirmed that the shortest paths weren't straight lines but a curved path, as Einstein's theory had predicted.

The bending and warping of space might also provide shortcuts across the universe, a way of getting around some limits implied by Einstein's theory of relativity. Einstein understood that the universe has a speed limit: the speed of light in a vacuum. Nothing can go faster than this. This causes a problem if you want to get from one side of the galaxy to the other. It's going to take time. This is a major problem faced by every science fiction writer: How do you get your cast of characters from one venue to the other without wasting years in transit? The secret often is to use a wormhole, a special solution to Einstein's field equations that provides a speculative shortcut between different bits of the geometry of space-time. A wormhole is a bit like a tunnel through a mountain, except this is a tunnel linking two points across the universe that ordinarily it would take you millions of years to travel between.

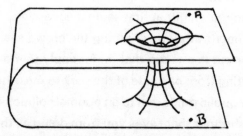

Figure 4.7. There's the long way around the universe from A to B or the shortcut through the wormhole

It turns out that Gauss's idea of light taking a shortcut along curved paths as it traveled from one hilltop in Göttingen to the other was correct. It's just that to see the effect he would have needed to work on a far grander scale, surveying not Hanover but our galaxy. To his credit, Einstein always acknowledged that the mathematicians of the nineteenth century had created the geometry that allowed him to discover relativity: "The importance of C. F. Gauss for the development of modern physical theory and especially for the mathematical fundament of the theory of relativity is overwhelming indeed. . . . I do not hesitate to confess that to a certain extent pleasure may be found by absorbing ourselves in questions of pure geometry."

SHORTCUT TO THE SHORTCUT

If you're planning a trip from A to B, then it's often worth considering the way light finds the quickest path: sometimes it pays to take a detour because the route is faster even if the path is longer. Measurements around the house can sometimes be tricky because you can't run a tape measure between two spots. But perhaps an angle is measurable. Sines and cosines were always meant to be fantastic shortcuts to measuring not just the night sky or the earth's surface but anything that might at first seem inaccessible. The urban planner's strategy of letting the crowd search out the shortcut is a tactic that might apply to more than just getting from one side of the park to the other. Letting the public guide you to an optimal solution is a potential shortcut that saves you from doing all the work on your own.

JOURNEYS

I LOVE WALKING. THE SLOW PACE allows me to experience the landscape and natural world in a way that often gets overlooked in our fast-paced lives. The walk is not about getting from A to B. Often it is about enjoying the long way around from A back to A. When my son was younger he found this absurd. We set out one day for a walk from a house we had rented in the countryside. After half a mile my son suddenly noticed a path breaking off from the one we were on. That path cut across a field, and on the other side of the field he saw our house. "Dad! I've found a shortcut! Look, we just need to take this path and we'll be home."

But for me walking is also a kind of shortcut. Three miles an hour seems to be the perfect speed for thinking. As Jean-Jacques Rousseau wrote in *Confessions*: "I can only meditate when I am walking, when I stop, I cease to think; my mind only works with my legs." Walking is my shortcut to mathematical revelation, the necessary detour that I need to make in order to allow my subconscious to explore a problem in a new way.

Robert Macfarlane talks about the connection between walking and thinking in his book *The Old Ways: A Journey on Foot*. He quotes Wittgenstein on a major breakthrough in his philosophical work he made while walking through the Norwegian countryside: "It seems to me that I had given birth to new thoughts within me." But it is the German word that Wittgenstein uses to describe those thoughts that Macfarlane points out is so revealing. The word Wittgenstein uses is

Denkbewegungen, which literally translates as "thought-ways." Macfarlane describes them as "ideas that have been brought into being by means of motion along a path (*Weg*)."

Macfarlane loves making journeys—being in the landscape, trekking and traveling. His books are a beautiful eulogy to the journey on foot. So I was really eager to talk with Macfarlane about his relationship to the idea of the shortcut. Is it possible that we might miss something by always looking for shortcuts?

"I can get the funicular railway up to the top of Cairn Gorm in my most beloved mountain range in northeast Scotland," Macfarlane told me, "and will feel that's the shortest cut to the top, but it will bring almost zero fulfillment and pleasure. But I'll come to the top of that same mountain over a two-day walk and it will be one of the most extraordinary places I've ever been."

Macfarlane also told me about the Scottish mystic and mountaineer W. H. Murray, whose writing captures the power of being in these places. "When his spirit is burdened or lightened the natural movement of a man's heart is to lift upward": Murray wrote these words on toilet paper he collected while incarcerated in a prisoner-of-war camp during World War II. Unable to journey with his body, in his mind he walked across the Scottish Highlands.

Another of Macfarlane's heroes is the modernist writer and poet Nan Shepherd. "Shepherd writes at the end of *The Living Mountain* in the 1940s about how these 'moments of being,' as she calls them, echoing Woolf and Wordsworth and others, are produced only when 'walking thus, hour after hour, the senses keyed, one walks the flesh transparent.' It's the most amazing phrase. 'These hills have no business with haste,' I think is how she puts it. So the shortcut is absolutely antithetical to revelation in that model."

But Macfarlane reminded me that many of the paths we walk for pleasure today were first trodden in Neolithic times because they were shortcuts. Living under conditions of scarcity, people had to balance energy expenditures, resources, and so on. They were not likely to pass up a shorter route if they found one, whether or not it offered the same kinds of contemplative opportunities as a longer one.

But not always. As Macfarlane points out, sometimes Neolithic cultures expended large amounts of resources in projects that weren't simply about the utility of survival. To this point, he tells me a beautiful story about hand axes that Neolithic people mined in Little Langdale, in Cumbria in the Lake District. "There were perfectly good hand-axe rock exposures at low levels in that valley, so they could have made use of those to get the tools that they wanted. But it's clear that they chose to ascend to much higher and more difficult ground on a crag called Gimmer Crag."

I was curious as to why they would go to the hard place to get the same rock that they could have gotten from the easy place. "There's an aura of place that remains with the object once the object's been detached from a place," Macfarlane noted. "So there are reasons why the 'long cuts' as well as shortcuts might have been taken prehistorically."

But then Macfarlane turned the tables on me, asking me if there were examples of the "long cut" in mathematics being unusually productive.

Conjectures are one example, I think. A conjecture is like the mountain peak. I don't want to look the answer up in the back of the book. That's like taking the funicular to the top of Cairn Gorm. The satisfaction of arrival depends on the days—nay, years—that it's taken me to reach that peak. But on the other hand I don't want to just slog through boring landscape for the sake of it. There are some walks that just feel like hard work.

There is a strange, delicate tension in mathematics between things being too easy that they become boring and things being so complex that understanding what is going on becomes impossible. John Cawelti, in his book *Adventure, Mystery and Romance*, characterizes the quality of this tension in literature, but it applies equally to mathematics: "If we seek order and security the result is likely to be boredom and sameness. But rejecting order for the sake of change and novelty brings danger and uncertainty. . . . [T]he history of culture can be interpreted as a dynamic tension between the quest for order and the flight from ennui."

Sometimes the fact that you have to take the long road to get to a peak is part of the joy. Fermat's Last Theorem went unproved for 350 years; it took generations of mathematicians taking journeys into strange esoteric lands before we found a way to reach the destination, the proof of the theorem. But those detours and "long cuts" are part of the joy of the proof. We discovered fascinating new mathematical lands that might have been left untouched if we hadn't been forced to travel around the mountain rather than tunnel through it.

It's interesting to consider whether, if the proof had been short and perhaps rather trivial, the value that we would have assigned Fermat's Last Theorem would have been much lower. Great unsolved conjectures, such as the Riemann hypothesis, get their aura from the challenge they offer and the work we need to put in to solve them. We talk about great conjectures being like climbing Everest. If it wasn't tough to reach the peak, perhaps we wouldn't value the achievement as much.

I tried to describe to Macfarlane that I think that what I enjoy in mathematics is not so much slogging across a moor but being stuck in front of a mountain, searching for a way through, and then feeling the extraordinary buzz of discovering a chink, a tunnel, the shortcut that gets me through.

But then Macfarlane interrupted me: "I'm watching your hands describe what you have to do, and what you look like was a climber. You look like a rock climber, not a walker. I'm talking here about gymnastic climbing, which is sort of distinct from mountaineering, which is distinct again from hill walking."

Was the challenge of rock climbing something Macfarlane enjoyed?

"I was very bad, but very into rock climbing for some years of my life, and climbers talk of the crux in a climb. Every great climb will have a crux move. It sounds as though that's very similar to the process you were describing where you'll work a problem. They're called bouldering problems. You'll start with the easy stuff and you'll do it again and again and you'll reach the crux and then you'll fall off. It'll chuck you and you can't quite make that dynamic leap. And then when you do it, the few times that I did it, it's absolutely thrilling. It is a problem-solving activity."

I certainly recognize that sense of frustration followed by elation that overcoming a mathematical bouldering problem can instill. Before our meeting I had just been watching *Free Solo*, a film that documents Alex Honnold's extraordinary scaling of El Capitan in Yosemite National Park without any ropes. The climb has about eight cruxes; it is the Riemann hypothesis of climbs. The hardest part of the climb is simply called the Boulder Problem; it's a difficult sequence across thin handholds, some no wider than a pencil, that are far apart. The wall is nearly vertical, and it requires a bizarre karate kick to get across. If he fails, he falls to his death. He doesn't have the luxury of being chucked off again and again. One of the things that struck me about the climb is that the shortest path to the top is definitely not a straight line. On his way up a peak Honnold often has to descend in the middle of the journey, moving away from the final destination for a time, in order to find a route that is climbable to the top. Geodesics in climbing are definitely quite strange lines that twist and turn up the mountain face.

I wondered what it is that determines which route a climber takes to the top of a mountain. Is it the route that's fastest? The most scenic? The toughest? There are eighteen named routes to the top of Everest, and several of them have never been climbed. The vast majority of climbers use one of two routes, called the South Col and the North Col. George Mallory, an English mountaineer, died trying to climb the North Col. He would talk of a "beautiful line." The beautiful line isn't necessarily the hardest route, but rather is the route celebrated for its beauty. This is interesting in light of the way mathematicians also talk about beautiful proofs. What is the quality that makes a route beautiful? As Macfarlane put it: "The beauty is typically a function of a kind of continuity of move-ment or line itself. So you don't necessarily need to traverse left and then pick up the next ridgeline or whatever. It will also be to do with the nature of the rock, if the rock is very non-friable and firm. It'll be literally just an elegance of the line you would draw in the air if you were de-scribing it. And then there's danger as well. The beautiful line unites all of those. And then there is a hardest line, the tiger line. And then there's what's called the diretissima, the most direct line." The term comes from the Italian climber Emilio Comici, who said: "I wish someday to make a route, and from the summit let fall a drop of water, and this is where my route will have gone." This same route also represents what's known as the fall line, the most perfect downhill gradient on a slope, the line water would take if it had a free run. This has been the key to some of the shortcuts that Macfarlane has taken to get off a mountain quickly when dangerous weather conditions or night might be closing in. "When you have to get off a mountain fast because you've got bad weather coming, or particularly in case you're going to be benighted, then you start to look for the fall line, because in theory that's the shortest cut to the lowest ground, which is probably where safety and shelter are going to be."

But you also have to consider the hazards that lie on the fall line. "The fall line might take you over a crag, and I know you don't want to do that. I can think of lots of occasions where I've had to move fast down and I'm busy balancing the fall line off against other hazard assessments that I'm carrying out. That's led me to some good decisions and some bad decisions. The shortcut can be a wonder but also a peril."

I asked if there had been any particular occasions when this shortcut came to the rescue.

"One of the best fall lines I've ever taken was when I surfed a small avalanche," Macfarlane said. "We were descending from a Scottish mountain and time was running late, and we came to a steep snow slope, and it was obviously ground that we wouldn't be able to cover if snow wasn't on it. But the snow kind of evens it out and solves some of the problems underfoot, and it's soft snow, like big granular sugar. So it wasn't going to avalanche in a big problematic way."

I must admit it sounded terrifying. Avalanches aren't generally things you want to encounter on a mountainside.

"We could see that it would run us out more or less safely down around 200 feet. So we just lay face-first on the slope and then let it take us. It deposited us, safe but wet, 200 vertical feet below where we started. It was brilliant. That was a good bit of risk assessment. That was one of the most exhilarating shortcuts I've ever taken."

CHAPTER 5

THE DIAGRAM SHORTCUT

Puzzle: Which song used in Quentin Tarantino's *Reservoir Dogs* is represented by the following diagram?

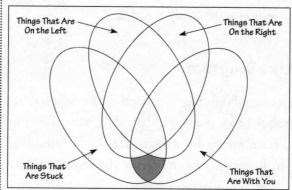

Figure 5.1. Venn that tune. Courtesy of Andrew Viner, author of *Venn That Tune*.

I**F A PICTURE IS,** as they say, worth a thousand words, then perhaps it is truly the ultimate shortcut. Leonardo da Vinci seemed to think so, anyway: "A poet would be overcome by sleep and hunger before being able to describe with words what a painter is able to depict in an instant." While the written word is a relatively new invention, humans have been developing the ability to interpret meaning in a visual image ever since we evolved as a species. Twitter revealed, for example, that tweets involving images or videos are three times more likely to be engaged with than those just containing text, which perhaps explains why more visually oriented social media apps such as Instagram are increasingly becoming the platform of choice for businesses trying to deliver content quickly and effectively. A well-designed image can be a remarkable shortcut, conveying your message much more effectively than the words you might use.

In mathematics too, sometimes a picture will communicate an idea where equations failed. The square root of −1 was for centuries regarded as a strange aberration by mathematicians. Ultimately it was Gauss's picture of imaginary numbers, depicting them as a 2-dimensional map, that brought these numbers into the mainstream. But it was shortly before Gauss's death in 1855 that the political power of a picture to represent numbers was truly revealed.

The Rose Diagram

When Florence Nightingale arrived at the hospital in Scutari, Turkey, in November 1854, she was horrified by what she found. The Crimean War had been raging for a year, and the hospital was responsible for tending to British troops injured in the conflict. The hospital was unhygienic, sordid, and overcrowded.

Immediately Nightingale went about improving conditions. She set up a laundry, brought in supplies, and provided nutritious food. But it didn't work—despite her best efforts, the death rate continued to rise. The sick and injured received assiduous care from Nightingale and other nurses, such as Mary Seacole, but nursing wasn't enough. Then, after she had spent months fighting this losing battle, two men arrived on the scene: Dr. John Sutherland, a cholera specialist, and Robert Rawlinson, a sanitary engineer. After some exploration, they found the underlying problem: the water system. It was choked with dead animals. And there was human excrement leaking from the privies into the water tanks. Rawlinson and Sutherland flushed the whole thing out. And then they started to see a difference.

As a result of the Sanitary Commission, as they were called, all the military hospitals rapidly improved. Within one month, deaths from infectious diseases had halved. Within a year, they had fallen by 98 percent, from well over 2,500 in January 1855 to 42 in January 1856.

After the war ended, Nightingale reflected on what she'd experienced over the previous eighteen months. She accepted that in war lives are lost during battle. But what she couldn't accept was the far higher number of deaths due to disease. She felt desperate about the losses—18,000 men had died, and many of them could have been saved. The challenge before her was how she could make lasting improvements in army hospitals so that the same tragedy would never happen again. But she knew that persuading the establishment of the urgency of radical reform wouldn't be easy.

Nightingale managed to secure an audience with Queen Victoria and her advisors. She impressed on them the need for an inquiry into why so many soldiers had died in these field hospitals. The Queen and the government were not anxious to pursue any further investigation into the war, but Nightingale's reputation was now legendary. So the government decided to ask her to write a confidential report that would be presented to a new Royal Commission. She wanted to help out, but what should she write? And, more important, how could she show the horror and tragedy that she'd seen unfold at Scutari?

Nightingale was afraid that the government would ignore her figures, so the essential facts, her clarion call to action, had to smack them in the eye. She created a diagram, now called the "rose diagram," to distill the message behind the numbers.

The diagram consisted of two roses. On the right-hand rose, she depicted the battle year 1854–55, with the soldiers' deaths shown for each month according to the cause of death. On the left, a smaller diagram shows the data for 1855–56. What matters is the area of each color. The central area, which she colored red (depicted in dark gray in our version), represented death from wounds; black indicated death from other causes, such as frostbite or accidents. But the staggering

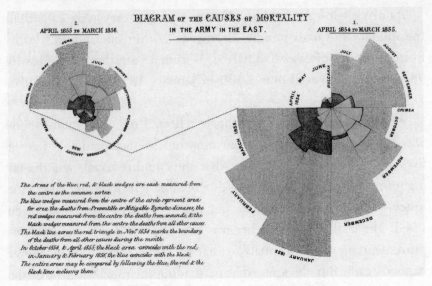

Figure 5.2. Florence Nightingale's rose diagram

number of deaths from infectious diseases such as dysentery and typhus appeared as large blue rose petals bursting out from the center (depicted in light gray here). As the winter of 1854 advances, the areas get steadily bigger, until in January 1855 more than 2,500 men are dying in a single month.

But the second rose shows that things didn't need to be like this. The far smaller blue region in the second diagram revealed that improved sanitation in the hospitals was the trigger for a dramatic drop in the numbers of lives lost due to infectious diseases.

It was this diagram, rather than all the words written in the report, that forced the British establishment to recognize that army medical practice was killing thousands of soldiers unnecessarily. Its strikingly visual appeal won over both hearts and minds, and set in motion a process of reform that would change medical care forever.

The diagram is meant to grab the eye first and engage the brain second. Nightingale wrote that the diagram should "affect thro' the Eyes what we fail to convey to the brains of the public through their word-proof ears." The diagram provides a shortcut to the message hidden in the numbers.

I learned of a more modern version of the power of visuals to persuade governments of health risks recently from Ian Lipkin, professor of epidemiology at Columbia University. He has been advising governments on responses to pandemics for many years. But he told me that his first attempt to explain to the US government the potential impact of a pandemic was met by stony silence. His thorough report, seven hundred pages long, had probably gone unread. So he prepared a highly condensed version. Still silence. Eventually he realized he needed to change the medium. Instead of words in a report, he helped to make a movie: *Contagion*, starring Matt Damon and Gwyneth Paltrow. The visual impact of seeing a virus kill so many people startled the US government into action, just as Nightingale's rose diagram had done with the British government in Victorian times.

Nightingale's rose diagram illustrates the power of representing a complex problem visually to provide a shortcut to understanding. It was not the first such diagram. In fact, she probably got some inspiration for the picture from the work of William Playfair. His *Commercial and Political Atlas*, published in 1786, contains forty-four images. Most of them are graphs plotting time against some other value, in the familiar *x/y* form. But one of the images is slightly different. It records exports and imports from Scotland not as a graph but as lines representing each of the values—a very early form of bar chart. This is the kind of diagram that Nightingale might have seen and considered herself.

Playfair believed that our brains had evolved to decode certain messages much more accurately when seen in pictures: "Of all the senses, the eye gives the liveliest and most accurate idea of whatever is susceptible of being represented to it; and when proportion between different quantities is the object then the eye has an incalculable superiority."

Today in this very visual age we are bombarded by pictorial representations of numbers. Diagrams to decode the secrets hidden in data are a powerful political and commercial tool. But as much as a good diagram can provide a shortcut to understanding, a bad one can completely mislead.

Certain news organizations are notorious for abusing diagrams to communicate a political message. Look at the following bar chart. The left-hand side was used to illustrate the apparently disastrous effect on taxes if George W. Bush's tax cuts expired. The difference looks enormous—until you suddenly notice that the vertical axis is starting not at 0 but at 34. Redraw the picture with the axis starting at 0, as on the right-hand side, and the difference is minuscule.

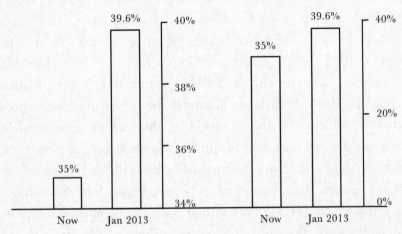

Figure 5.3. Two different perspectives on the effect of tax cuts

Another classic misuse of bar charts is the following:

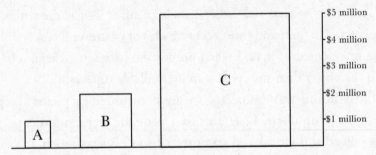

Figure 5.4. A misleading diagram of company profits

The graph is meant to show the dominance of Company C over Company B and Company A. It is just the height of the icon that is important in recording the data. Yet by also scaling up the width, the company has completely exaggerated its importance. Although Company C's sales are 5 times bigger than Company A's, the icon for Company C is 25 times the size of Company A's icon.

In some ways Nightingale's rose diagram misses a trick here. She created it so that the rose's area corresponded to the numbers, but because the petal area expands in every direction, she ended up downgrading the impact. If she had replaced the rose with a bar chart, the heights of the sections corresponding to the blue regions would have been even starker in contrast to the rest.

Mapmaking

A map is probably the perfect example of a diagrammatic shortcut. It is not a replica of the land it is mapping. For a start, the whole point is that a map is a scaled-down version of the land being surveyed. Even then, many features have to be thrown away. But map the landscape well, choosing the essential features to include and throwing away what is not necessary, and you have an amazing shortcut to finding your way.

I've always enjoyed the story that Lewis Carroll tells in his last novel, *Sylvie and Bruno Concluded*, of a country that didn't appreciate the importance of discarding information when making maps. They prided themselves on how accurate their maps were:

"We actually made a map of the country, on the scale of a mile to the mile."

"Have you used it much?"

"It has never been spread out, yet. The farmers objected: they said it would cover the whole country, and shut out the sunlight! So we now use the country itself, as its own map, and I assure you it does nearly as well."

As Carroll humorously points out, maps need to make choices about what to leave out.

Some of the first maps made by humans are maps of the heavens, not the earth. The drawings on the cave walls at Lascaux map the configurations of the Pleiades, stars that were often used to mark the beginning of a cycle of the year. One of the first maps of the earth is a clay tablet carved by a Babylonian scribe maybe as long ago as 2500 BCE. It shows a river valley between two hills. Hills are depicted by semicircles, rivers by lines, and cities by circles. It also includes directions for which way to orient the map.

The Babylonians are responsible as well for the first attempt to map the whole world, which dates to 600 BCE. The map is very much a symbolic rather than literal attempt. It depicts a circular shape surrounded by water, the Babylonian view of the lay of the land.

But once it was recognized that the earth was spherical rather than flat, creating a flat map of a spherical surface became an interesting challenge for cartographers. It is the sixteenth-century Dutch cartographer Gerardus Mercator who is generally credited with finding a clever solution.

Because this is the age of the exploration of the planet by sea, Mercator's principal goal was to create a map that would help sailors get from one point on the planet to another. The principal tool for navigation was a compass. The easiest way to get from A to B was to know a fixed compass direction to head in, such that if the boat was kept at that angle, then you would arrive at your destination.

Such lines have constant angle to the lines of longitude running north to south. They are called rhumb lines, and drawn on a globe, they can be seen to spiral into the North Pole.

They are not the shortest paths from A to B, but if you are more concerned about not being thrown off course, these were by far the best paths to take.

Mercator's map has the wonderful property that these curved paths get turned into straight lines on the map. If you wanted to find the angle to set course to get from A to B, you just had to draw a straight line on

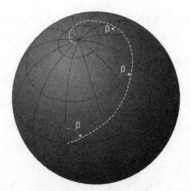

Figure 5.5. A rhumb line maintains a constant angle
to the lines of longitude

Mercator's map and the angle of the line to the lines of longitude running north would be the angle you'd need to maintain as you sailed
across the ocean.

This projection of the sphere onto the rectangle is called a confor-
mal mapping because it preserves angles. It can be achieved by doing
the following. Imagine the earth is a balloon, with wet ink all over its
surface. Wrap a cylinder around the earth so that the cylinder touches
the equator. Now start to inflate the earth, so that gradually the surface
of the globe comes more and more into contact with the cylinder, the
ink printing out a map of the surface. Unwrap the cylinder and you
have your map.

Obviously, I can't map the poles onto the map in this way, so the
top of the map will be a line of latitude close to the pole. The effect of
this map is to stretch out the lines of latitude as we head north or south
from the equator.

The map was a fantastic tool for those at sea. This was obviously
Mercator's aim, given that he gave the map the title "New and more
complete representation of the terrestrial globe properly adapted for
use in navigation."

Whereas angles between lines on the globe were preserved in the
map, areas and distance were not. And this has had huge political implications. Because of how useful the map was for centuries, it became

the accepted view of what the planet looked like. But one of the effects of the map is to greatly inflate the importance of countries that are far from the equator, countries such as the Netherlands and Great Britain. For example, draw a circle on the equator and a circle of the same size on the surface of the sphere located over Greenland. This second circle grows ten times in size when it is mapped onto Mercator's projection. Africa, for example, is made to look about the same size as Greenland when in reality it is fourteen times larger.

The Mercator map fell afoul of postcolonial politics, and an alternative map called the Gall-Peters map was adopted by UNESCO. It is the map widely used in British schools, but many school districts in the United States still have not replaced the Mercator map (Boston's schools did so, but only in 2017). For many US citizens, the shrinking in size of the United States seen with the Gall-Peters map does not mesh well with their citizens' view of the place of their country in the world.

The truth is that any map is going to make compromises. This was in fact a discovery that Gauss made when he was investigating the nature of curvature of different geometries. In a discovery he called his Theorema Egregium (Remarkable Theorem), he proved that a flat map cannot be wrapped around a spherical globe without distances being distorted. Something has to be compromised in any map of the planet. In the Gall-Peters map, the areas may be right, but the shapes of the continents are not. Africa looks twice as long as it is wide, while in reality it is more square-shaped.

Of course, there is another major choice that is made by most maps, which is to put the Northern Hemisphere at the top and the Southern Hemisphere at the bottom. But because a sphere is symmetrical, there is no reason not to have had the map the other way up. Again, the choice reflects that it was residents of the Northern Hemisphere who were drawing up the maps.

Australian resident Stuart McArthur decided to counter this northern bias in maps by producing a map with the Southern Hemisphere on the top. When you see it for the first time, it is quite a shock. It just doesn't look right. And yet this is just a reflection of how much we've gotten used to Mercator's vision of the planet.

Maps are all about what it is you are trying to achieve. Is this a shortcut to navigation? A shortcut to understanding land size? Most maps try to preserve some geometric feature. Perhaps distances on the map correspond to distances on the planet. Or angles between lines are the same. But sometimes a good map throws all these things away and just preserves the most important features of how to get from A to B.

One of the maps I love most and use on a daily basis is the map of the London Underground. A physical map showing the geographic locations and routes on the London Underground is not a very helpful picture if you're trying to negotiate your way around the city. Instead, Harry Beck's iconic map, developed in 1933, isolates the way the parts of the network are connected together while ignoring physical dimensions. The map was so revolutionary that initially it was rejected by the company running the Underground. But the trouble was that they were losing money because Londoners were not using the system. When they tried to find out why people weren't using it, they discovered that people just couldn't navigate around the network. The maps they had produced were trying to replicate the geography of the city, but that resulted in a rather cramped, tangled mess of lines that people found difficult to read.

Beck had seen the problem and had decided that he needed to abandon geographical accuracy. Instead he pushed and pulled the lines around, straightening them out, making them cross at clean angles, stretching stations apart. It may have helped that Beck had a background in electronics, because the map resembles something closer to the layout of an electronic circuit board rather than a train map.

Having realized that they needed a better map that passengers could use to navigate the system, the company eventually decided to adopt Beck's proposal, and 750,000 copies of the map were printed and distributed. The map is now an international icon. It's inspired works of art: on Simon Patterson's reworking of the map, which hangs in the Tate Modern in London, station names are replaced by the names of engineers, philosophers, explorers, planets, journalists, soccer players, musicians, film actors, saints, Italian artists, scholars of China, comedians, and "Louis" (French kings). J. K. Rowling, author of the Harry Potter books, even gives Professor Dumbledore a scar on his left knee in the shape of the map, a nod to the fact that she had her best ideas for the series of books while sitting on trains.

The power of the London Underground map is that it is not a geographic map but focuses instead on the more important quality of how to get from A to B. The fact that the same length of line is used to represent the connection between Covent Garden and Leicester Square as between Kings Cross and Caledonian Road does not mean the distances are the same. For a commuter, knowing there is such a connection is much more important than knowing the distance between stations.

This is an example of a new way of looking at the world that was introduced in the mid-nineteenth century. Often the exact distances between objects are not important. It is how they are connected together that is key to the identity of the shape. This new way of looking at the world is called topology. Gauss was one of the first to start considering how the properties of surfaces might depend less on their physical geometry and more on how points on the surface were connected. Although he never published his ideas, they were the inspiration for the work of Johann Benedict Listing, who first used the term *topology* in a paper in 1847. We shall see in Chapter 9 how topological maps can be a very handy shortcut to finding your way around networks, and not just on the London Underground.

But diagrams need not be restricted to showing the physical connections between locations in London. Very effective use has been made of maps that replace Underground stations with mental ideas. Called mind maps, their purpose is to tease out interesting connections between different ideas that you might be exploring. Mind maps have for years been the staple diet of students trying to cram for exams because they help to create an integrated story of a subject that in words can feel too difficult to navigate. In some ways they are tapping into Ed Cooke's memory palaces. The mind map can turn a hodgepodge of ideas into a physical journey you can navigate on the page.

These diagrams have a long history. Newton made doodles in his notebooks showing a sort of mind map he used as an undergraduate in Cambridge to reveal his thoughts on how different philosophical questions might be interrelated. The point is that these maps hope to disrupt the rather linear manner in which a textbook might present ideas and instead try to mimic the more multidimensional manner in which we process ideas in our minds.

Mapping the Big and the Small

As Leonardo articulated, the visual world can describe things that will forever transcend the written word. A single image can convey the simple underlying pattern hidden by the complexity of words or equations. But a diagram is more than just a physical representation of what we see with our eyes. The power of a diagram is to crystallize a new way of seeing the world. Often it requires throwing away information and focusing on what is essential, as Lewis Carroll's humorous unscaled map illustrated. Other times it changes a scientific idea into a visual language, providing a new map where the mathematics of geometry takes over and helps us to navigate the science at hand.

Copernicus certainly understood the power of a good picture. In his great opus *De revolutionibus orbium coelestium*, published shortly before his death in 1543, Copernicus takes 405 pages of words, numbers, and equations to explain his heliocentric theory. But it is the diagram that he draws at the beginning of the book that captures in a simple image his revolutionary new idea: it is the sun that is at the center of the solar system, not the earth.

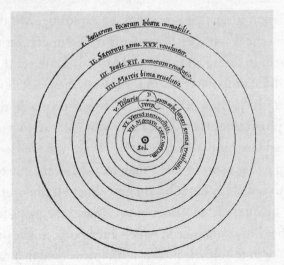

Figure 5.6. Copernicus's diagram of a sun-centered solar system

His picture encapsulates some of the essential elements of the best diagrams. The concentric circles are not meant to describe the precise orbits of the planets. Copernicus knew they weren't circles. The uniform distances between the circles aren't meant to tell you how far the planets are from the sun. Rather, this picture conveys the simple yet shocking idea that we aren't at the center of things. His diagram transformed our view of our place in the universe.

Today modern cosmologists use diagrams to chart the entire universe over its 13.8 billion years of history, diagrams to capture the workings of massive black holes, diagrams to navigate the complexities of 4-dimensional space-time. The power of a diagram to provide a shortcut to the vastness of the universe is perhaps the only way we could manage to conceive of our place in something that at first sight seems inconceivably large.

But diagrams can also work like magnifying glasses, allowing us to see the very small. Go into any chemistry lab and you will see chalked on the blackboards letters connected by single, double, and sometimes triple lines. These are the diagrams that tell chemists how atoms are put together to make up the molecular world.

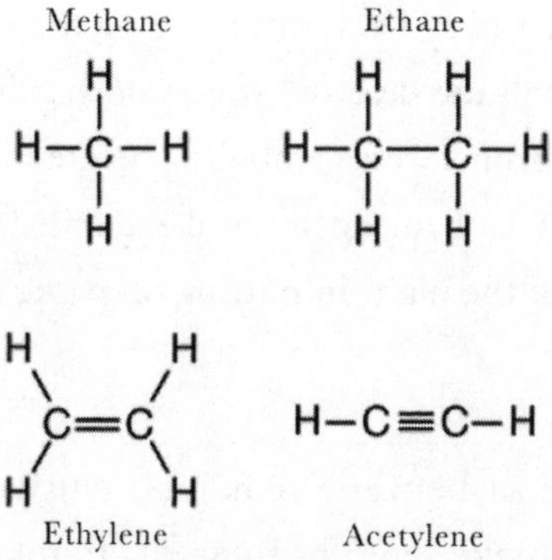

Figure 5.7. Molecular diagrams

A picture of methane shows a central C with four lines emerging out of it and Hs at each end to depict a molecule of CH_4. The colorless flammable gas ethylene C_2H_4 has a slightly different structure, with a double line between the Cs and four Hs attached to the Cs. How molecules might react and change can be navigated by using these diagrams. The presence of a double bond is often accompanied by the molecule being reactive in the presence of a substance that likes electrons. In chemistry we get so used to manipulating these diagrams that we forget they are a shortcut to the extraordinary reactions that are happening on a scale that is hard for a microscope to pick up. But these diagrams can also lead to the discovery of new structures hidden inside the molecular world.

As illustrated by the methane molecule, carbon likes to have four lines emerging from it. Hydrogen will only have one line. So it was something of a mystery when the benzene molecule was first extracted by Michael Faraday in 1825, analyzed, and found to be made up of six carbon atoms and six hydrogen atoms. If you try to make a diagram of its structure the numbers just don't seem to work out. It seems impossible to cover the six greedy carbon atoms, each of which has four arms emerging from it, with just six single-armed hydrogen atoms. It was August Kekulé, a German chemist working in London, who finally cracked the mystery:

One fine summer evening I was returning by the last bus riding outside as usual, through the deserted streets of the city. . . . I fell into a reverie, and lo, the atoms were gamboling before my eyes. . . . The cry of the conductor: "Clapham Road" awakened me from my dreaming; but I spent a part of the night in putting on paper at least sketches of these dream forms.

Still the structure of benzene remained elusive. He worked many late nights trying to make sense of these diagrams, but it took another dream for the secret to finally reveal itself:

I turned my chair to the fire and dozed. Again the atoms were gamboling before my eyes . . . long rows sometimes more closely fitted together all twining and twisting in snake-like motion. But look! What was that? One of the snakes had seized hold of its own tail, and the form whirled mockingly before my eyes. As if by a flash of lightning I awoke.

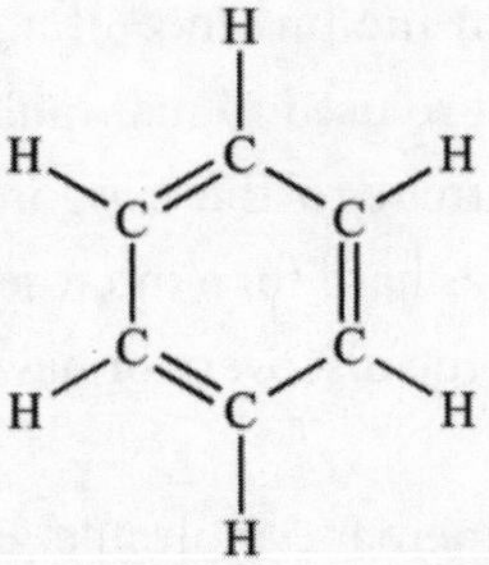

Figure 5.8. The ring structure of benzene

He had it. The carbon atoms weren't a long line but a ring. This was the way to use up the carbon arms. They were shaking hands with each other, meaning that just one arm each was available to shake the hand of a hydrogen atom. The discovery of the benzene ring and similar ring structures in other molecules would guide the development of a new field of chemistry. It turns out that many molecules with such ring structures are aromatic. For example, if you swap one of the hydrogen atoms for a molecule of COH, then the resulting molecule smells of almonds. Change that to a slightly longer molecule of C_3OH_2 and the smell changes to cinnamon.

These molecules are simple enough that their structure can be captured by a 2-D diagram. But more complex molecules, such as hemoglobin, are much more challenging to capture in a picture. Indeed, it took the work of an artist to fully capture the complexities of a molecule of hemoglobin for the first time. Irving Geis had trained as an architect and fine artist, but it was his extraordinary picture of the architecture of hemoglobin that appeared in *Scientific American* in 1961 for which he became famous.

John Kendrow had successfully pieced together the structure of the protein by using a large number of 2-D X-rays, work for which he received the Nobel Prize in 1962. It was an extraordinary feat. The molecule consists of more than 2600 atoms, and that is still quite small for a protein molecule. Although he'd managed to make a picture of the structure in 1957, he decided he needed the help of a master draftsperson to really capture his discovery. That's when he turned to Geis. After six months of work poring over Kendrow's papers and models, he produced a watercolor image that appeared in the June 1961 edition of *Scientific American*. It is a stunning image, but almost too complex to provide a shortcut to really navigating the molecule's properties.

Probably the ultimate challenge came with trying to picture DNA. As I've stressed, the power of a good diagram often emerges when we throw away information. When Francis Crick and James Watson discovered the structure of DNA with its double helix, they could have drawn an incredibly complicated image in their paper in *Nature*, with the full molecular description. But the essence of their discovery was the two strands that make up DNA and which explain how the molecule enables our genes to be passed down the generations. They famously announced the discovery in the pub in Cambridge where they used to drink. When Crick rushed home to say he'd found the secret of life, his wife was rather dismissive: "He was always coming home and saying things like that."

Interestingly, his wife, Odile, would play a major role in bringing the discovery to the attention of the world, because she created the diagram that appeared in their *Nature* paper. Crick had given her a sketch of the sort of thing he wanted, but he didn't have the artistic skills to reveal the important message in their discovery. Odile was a trained professional artist. She'd studied in Vienna in the 1930s and had then gone on to St. Martin's in London and the Royal College of Art. She did occasional portraits of her husband, but most of her work focused on the female nude. Molecular structures weren't really her thing.

But as Francis explained the discovery with the aid of his rather cluttered sketch, Odile got the point and turned his vague impressions into a memorable image, the power of which she probably didn't realize at the time. This double helix has become a symbol that goes far beyond just DNA, biology, or even scientific discovery.

From the beginning, the double helix appealed to artists. One painter who was quick to add it to his palette of scientific metaphors was the surrealist Salvador Dalí. He called this his period of "nuclear mysticism," and his use of DNA revealed a surprisingly conservative and religious aspect to his art.

But for me one of the most amazing uses of diagrams is the Feynman diagram. It not only enabled us to see things that were impossible even for a microscope to pick up, but also allowed us to shortcut the need for some extraordinarily complex calculations.

While a chemist's blackboard is covered in Cs, Hs, and Os connected together with lines, on the physicist's blackboard you will probably find diagrams representing the interactions of the fundamental particles that make up the chemist's atoms. These are dynamic diagrams that show the evolution over time of what might happen when, for example, an electron and a positron interact.

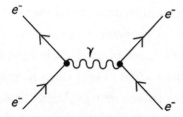

Figure 5.9. Feynman diagram of the interaction between
an electron and a positron

The diagrams are an amazing bookkeeping device dreamed up by the physicist Richard Feynman to keep track of hugely complex calculations of mathematical integrals. He first revealed his discovery of this diagrammatic shortcut to particle interactions at a closed meeting of twenty-eight leading theoretical physicists who had gathered to try to make sense of the calculations that the theory of quantum electrodynamics (QED) was throwing out. The meeting took place at the Pocono Manor Inn in rural Pennsylvania during the spring of 1948.

A young wunderkind from Harvard, Julian Schwinger, had spent the day explaining his complex mathematical approach to QED. It was a marathon all-day lecture punctured with a few breaks for coffee and lunch. The audience sat glued to their seats as the mathematics bubbled up on the board explaining the interactions. By the end of the epic lecture, their brains were probably fried. That perhaps explains why when Feynman then stood up at the end of the day to explain his approach and started drawing pictures on the board, people were initially perplexed by how these diagrams helped to do calculations. Indeed, some of the big guns who sat through the lecture, such as Paul Dirac and Niels Bohr, were flummoxed by Feynman's pictures, and concluded that the young American simply did not understand quantum mechanics.

Feynman left the meeting disappointed and depressed. But the diagrams were eventually rescued by one of the other big names in physics, Freeman Dyson, who understood that they were in fact equivalent to the complex mathematical calculations that Schwinger was doing. It wasn't until Dyson explained the insight during a lecture that the physics community started to take these pictures seriously. The articles that Dyson wrote subsequently offered a how-to guide to the diagrams, including step-by-step instructions for how the diagrams should be drawn and how they were to be translated into their associated mathematical expressions.

Today it is these diagrams concocted by Feynman that are the first port of call for any theoretical physicist trying to tease out what happens when particles interact with each other. They are an amazing diagrammatic shortcut to the fundamental interactions happening at the foundations of the physical universe. No experiment has ever captured a quark in isolation, and yet on the blackboard these diagrams give us a way to navigate the evolution of a quark as it interacts with its environment.

My Oxford colleague Roger Penrose has developed a similarly powerful visual shortcut to some of the most complex ideas in fundamental physics. His theory of twistors, proposed in 1967, is an attempt to unify quantum physics, the physics of the very small, with gravity, a physics generally of the very large. It is a hugely mathematical theory, but for Penrose, the best way to navigate the complex mathematics was by drawing pictures. Fortunately, Penrose is quite an adept artist in his own right, having had some interesting interactions with the Dutch visual artist M. C. Escher. His artistic skills probably helped him create the diagrams, which are the best shortcut to coping with the complex mathematics.

Although introduced in the late sixties, Penrose's ideas have become mainstream thanks to new work connecting his theory with current ideas. Recently one of the diagrams that have emerged out of this new approach, the amplituhedron, which represents an eight-gluon particle interaction, has provided an amazing shortcut to understanding the physics. Using Feynman diagrams, the same calculation would have taken roughly five hundred pages of algebra.

"The degree of efficiency is mind-boggling," commented Jacob Bourjaily, a theoretical physicist at Harvard University and one of the researchers who developed the new idea. "You can easily do, on paper, computations that were infeasible even with a computer before."

Venn Diagrams

You might recognize the sort of diagram that was used in the puzzle I set out at the beginning of the chapter. Those who learned math during the "new math" revolution will probably have used Venn diagrams often as a way to organize information. Each circle represents a concept, and the regions where the circles do or do not intersect give all the different logical possibilities for how these concepts relate to each other. For example, take the idea of a number being (a) a prime number, (b) a Fibonacci number, and (c) an even number. We can distribute the numbers from 1 to 21 according to how many of these categories they satisfy.

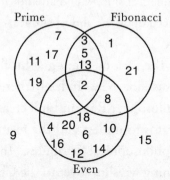

Figure 5.10. The Venn diagram of prime numbers, Fibonacci numbers, and even numbers

The Venn diagram is a clever pictorial way to represent the different possibilities. For example, it reveals that 2 is the only even prime (mathematicians think it's funny to say 2 is an odd prime, because it's the only even prime). There are no numbers that are even and prime but not Fibonacci numbers.

This type of diagram is named after the English mathematician John Venn, who in 1880 introduced them in a paper entitled "On the Diagrammatic and Mechanical Representation of Propositions and Reasonings." The diagrams were meant to help navigate the logical language that his contemporary George Boole was developing. As well as diagrams, Venn specialized in making bowling machines for cricketers to practice their batting with. The Australian cricket team asked to try out the machine on a visit they made to Cambridge, where Venn worked. They were rather shocked when the machine successfully bowled out their captain four times in a row. But Venn thought his diagrams were of more lasting importance:

> I began at once somewhat more steady work on the subjects and books which I should have to lecture on. I now first hit upon the diagrammatical device of representing propositions by inclusive and exclusive circles. Of course the device was not new then, but it was so obviously representative of the way in which any one, who approached the subject from the mathematical side, would attempt to visualize propositions, that it was forced upon me almost at once.

He was right that the idea of using graphical images to represent logical possibilities was not new. Indeed, there is evidence of the thirteenth-century philosopher Ramon Llull creating something similar. He used his diagrams to understand the relationship between different religious and philosophical attributes. They were meant as a debating tool for winning Muslims over to the Christian faith through logic and reason.

But it's Venn's name that has stuck. Most often you will see the diagrams considering just three different categories. That's because this seems the easiest diagram to draw on paper to represent all the possibilities. When you get to four different categories, you have to work much harder to make your four regions intersect in such a way that all logical possibilities are covered. For example, this isn't good enough:

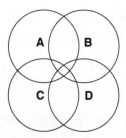

Figure 5.11. This diagram does not represent the Venn
diagram of four sets

There is nowhere that represents being in region A and region D
but not in the other two. Instead you need a diagram that looks like this:

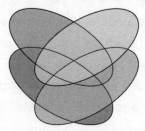

Figure 5.12. The Venn diagram of four sets

A seven-set Venn diagram, though, is starting to miss the point
about diagrams being aids to understanding:

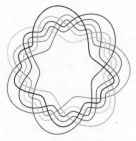

Figure 5.13. The Venn diagram of seven sets

One of my favorite books has been *Venn That Tune* by Andrew
Viner, who used Venn diagrams to capture song titles. The challenge at

the beginning of this chapter is one of his. The diagram is a shortcut for "Stuck in the Middle with You" by the British folk-rock band Stealers Wheel.

SHORTCUT TO THE SHORTCUT

How would you depict your message or data as a picture or diagram? There are many different genres you can employ that might provide a shortcut to understanding: A simple graph showing your business's profits at different times of year. A bar chart keeping track of the most popular dishes in a café. A Venn diagram to explain overlaps and differences in opinions between different political parties. Or perhaps a network diagram, like the London Underground map, to reveal connections between ideas that words are obscuring.

PIT STOP 5

ECONOMICS

"THE MOST POWERFUL TOOL IN economics is not money, nor even algebra. It is a pencil. Because with a pencil you can redraw the world." These are the opening lines of Kate Raworth's book *Doughnut Economics*, in which she explains a new diagram that challenges the twentieth-century story of economics. It is a diagram in the shape of a doughnut.

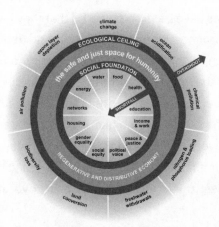

Figure 5.14. The doughnut economic diagram

I am a big fan of Raworth's book, partly because the doughnut (or torus, as we call the shape in mathematics) is one of my all-time favorite shapes. Not just because it is delicious to eat but because the mathematics of this shape is fascinating. Understanding its arithmetic is at the heart of proving Fermat's Last Theorem. The topology of this shape is crucial to understanding the possible shape of the universe. But as I discovered in Raworth's book, it is also key to a revolution in economics. So I was keen to talk to her about the genesis of this game-changing diagram as a shortcut to economic thinking.

Open any economics book, attend any economics lecture, watch any economics video, and you'll invariably find a couple of diagrams that appear over and over again. One is a picture of growth, which always shows a graph bending upward in an exponential pattern, promising a future of seemingly unlimited production. The other image is a graph with two curves crossing in an X shape, depicting supply and demand plotted against quantity and price. Demand shows how the cheaper the price, the more someone can buy. Supply shows that the amount a supplier will produce goes up if the price goes up. Putting one on top of the other is meant to reveal the economic equilibrium, when the price at which quantity demanded equals the quantity supplied.

These diagrams have been so powerful that they have led to the idea that economics is, at heart, only about supply and demand. But Raworth wanted to challenge this twentieth-century economic model. It was missing so much that was important to understanding global economics: the environment and human rights. As George Monbiot wrote in his book *Out of the Wreckage*, the best way to counter one story is with another story. Raworth's philosophy is similar: "These old diagrams are like intellectual graffiti on the mind, and like graffiti, it's very hard to scrub out. The best thing you can do is paint over it with something new."

Raworth has always found visuals the best way to understand complexity. "At school I was discouraged from drawing pictures in the margins of books, but now we understand that there are many forms of intelligence and visual intelligence is one of them. I loved reading Feynman as a teenager. His books were full of drawings. Maybe that told me early on that this is part of understanding even though other people said I was doodling."

Raworth went on to study economics but felt that the subject didn't really understand how human societies work. "I began to really be ashamed of the concepts I'd been taught," she said.

It was while on jury duty that Raworth came across a diagram by Herman Daly, an economist working at the World Bank, that sowed the seed for her economic insight. Daly wanted to challenge the assumption of limitless growth and so proposed that an outer circle be drawn around the economists' graphs that should be labeled "the environment."

"The power of a great diagram," Raworth said, "is that once you've seen it, you can't unsee it. It allows you to make a mindset leap, a paradigm change."

For years, Daly's diagram sat in the back of Raworth's mind until one day, when she was working at Oxfam, a diagram inspired by Daly's idea sparked her diagrammatic epiphany. That diagram changed her perspective on economics—and, she hopes, the world's. It was Johan Rockström's diagram of the nine planetary boundaries that represent the safe operating space for humanity. There was Daly's circle but now there were big red regions radiating out from the middle, each region representing the current status of things such as the ozone layer, the water cycle, climate, and ocean acidity. The trouble was that many of them were overshooting the circle.

"I just had this visceral reaction," Raworth recalled. "This, I realized, is the beginning of twenty-first-century economics."

But it wasn't just a pretty picture. It was backed up by numbers. Economists usually measure everything by converting it into dollars, going back to the common currency of utility of Jeremy Bentham and John Stuart Mill. This is meant to be a clever shortcut allowing them to compare seemingly incompatible quantities. One number to rule them all.

But that number doesn't tell us everything we need to know. Raworth asked me to imagine a car: "I've made it really simple. There's just one dial, and I've added into that dial speed, temperature, revs, how much gas is left in the tank, because I don't want to bother you with all this different information. You would never drive that car. You actually want the dashboard. Humans are very good at dashboards. We fly planes with complex dashboards. We live in a complex system. Hiding the complexity isn't going to give you a richer decision tool. It's a dangerous shortcut."

That was what made these new charts exciting. They weren't using dollars as the sole metric. Instead, they used multiple metrics: metric tons of carbon dioxide, tons of fertilizer use, ozone metrics. And yet Raworth believed the diagram was still missing an important component: the human.

"I was sitting in Oxfam surrounded by people who were responding to an emergency drought in the Sahara or campaigning for health and education for kids in India and thinking, 'Well, if there is an outer circle representing a limit of pressure that humanity could put on the planet, then there's also an inner limit which we've called for nearly seventy years human rights. The right of how much food everybody needs each day or how much water you need or the minimum conditions of housing or education to be part of society. If there's this outer circle, then I decided we needed to draw in an inner circle."

At this point Raworth went up to the whiteboard in my office and drew a doodle of a doughnut with an outer ring for the environment and an inner ring for human rights.

Initially, Raworth kept the diagram to herself. Then, at a meeting of Earth system scientists in 2011 discussing the nine planetary boundaries, someone turned to Raworth, who was there as the representative from Oxfam, and said: "The problem with this framework of planetary boundaries is there's no people in it."

"There was a big whiteboard on the wall. I said, 'Can I draw a picture?'" She jumped up and drew the doughnut on the whiteboard, as she had done in my office, and then explained that just as we need an outer circle to bound the impact humans are having on our environment, we also need an inner circle that represents the minimum conditions for every human on our planet in terms of food, water, healthcare, education, and housing.

"We need to use Earth's resources to meet everyone's needs, but not use them so much that we go beyond the limits of the planet. We want to be in this space in between," she told me, pointing to the doughnut. "I drew really quickly because I thought they'd say, 'Yes, dear, sit down.' But instead they responded excitedly that that's the picture we've been missing all along, and it's not a circle, it's a doughnut."

Raworth wrote up her diagram as a discussion paper for Oxfam and published it, to immediate enthusiasm. "That was the moment when I was really struck by the power of images as a shortcut. If you take all the words that were in it—food, water, jobs, income, education, political voice, gender equality, climate change, ocean acidification, ozone layer depletion, biodiversity loss, chemical pollution—and wrote those things in a list, nobody would blink. But if you draw them in relation to each other in a pair of concentric circles, people say this is a paradigm change."

As John Berger wrote in his classic 1972 book *Ways of Seeing*: "Seeing comes before words. The child looks and recognizes before it speaks."

For Raworth, a diagram is a shortcut, but it also encapsulates a worldview. And this is its danger, because it might simply be a shortcut to the way *you* see the world. It may in fact hide things that in your view are not important but that for others might be fundamental to their vision. If a company is only interested in short-term corporate profit, it might be happy with the exponential growth graph, but if you care about the environment, then hiding the impact of growth on climate means the shortcut is very selective in who gets quickly to their desired destination. It has put another group far from the goal they are seeking. Since a diagram throws away extraneous data, it borders on cutting corners. The corners you are cutting off, Raworth believes, are potentially a reflection of your worldview. What is a shortcut for one economist to explain their worldview might be completely the wrong path according to another, leading people away from what they believe is the correct destination.

"The shortcut might be leading you down an extremely dangerous hole. The quote I've come to love is from a mathematician called George Box: all models are wrong, but some are useful."

The doughnut became one of seven new diagrams that Raworth offers in her book *Doughnut Economics* as a shortcut to a new economic destination. She admits that, just like digging a tunnel through a mountain, creating these shortcuts was hard work. But it was work that is urgent given the direction toward which the planet and humanity are heading.

"To rewrite economics so it's a fit tool for the twenty-first century, we need to use all the shortcuts we can, because we have little time!"

CHAPTER 6

THE CALCULUS SHORTCUT

Puzzle: Which is the shortcut? If you rolled balls
down each of these slopes, which one would
get the ball to the end fastest?

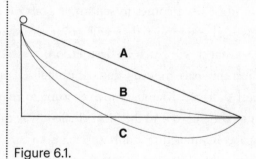

Figure 6.1.

A S ASTRONAUT JOHN GLENN orbited earth for the third time, he
began to prepare the spacecraft for its reentry into the earth's at-
mosphere. It was February 20, 1962, and Glenn had just become the
first American to orbit the Earth. But he still needed to get home safely
in order for the mission to be counted a success. The trajectory he was
going to choose would be crucial. Get the angle of descent wrong and
the spacecraft would burn up on entry. Land the craft too far out to sea
and the navy would not get there in time to stop the capsule from
plummeting to the seabed.

Glenn was putting his life in the hands of the calculators that had
crunched the numbers. In 1962 these calculators were not machines.

They were a group of women now immortalized in the Hollywood film *Hidden Figures*. In the film Glenn sits on the launchpad before initiating blast-off and says to mission control, "Get the girl to check the numbers." The "girl" in question was Katherine Johnson, one of the team of calculators NASA was using. In the movie it takes her twenty-five seconds to run the math and confirm that everything is on course.

In reality, the calculations Johnson did occurred weeks before blast-off and probably took two to three days. Even so, that's still an impressively fast time to navigate such a complex range of possible paths and scenarios. But Johnson had a shortcut up her sleeve that allowed NASA and every agency that has ever sent objects into space to know where their spacecraft would end up: calculus, probably the most powerful tool for finding shortcuts ever invented by mathematicians. From landing a probe on a comet to sending a craft on a flyby of the planets, calculus is the signpost that will point the spaceship in the right direction so that it will reach its destination.

And it's not just the space industry that has capitalized on the power of this mathematical shortcut. Many companies want to maximize output while minimizing cost and to find the most efficient way to produce their product. Aerospace manufacturers aim to create an aircraft wing that causes the least amount of drag so that fuel isn't wasted. Tankers need to find the fastest route through turbulent waters. Brokers are trying to spot the moment a stock hits its maximum value before it crashes. Architects want to design buildings that maximize space given the constraints of the surrounding environment. Engineers constructing bridges need to minimize materials without compromising structural stability. All of these professions need calculus to achieve their goals. If you've got a complicated equation describing the economy or energy consumption or whatever you're interested in, then the calculus is a way of analyzing that equation and finding the points where the output of the function is greatest or least.

It's also a tool that gave the scientists of the seventeenth century the ability to understand a world in flux. Apples were falling. Planets were orbiting. Fluids were flowing. Gases were swirling. Scientists wanted a way to take a snapshot of all these dynamics scenarios. And calculus provided a way to freeze-frame all this motion. It is striking that it mirrored the interests of artists working at the same time: baroque painters depicted soldiers falling from horses, architects designed buildings with sweeping dynamic curves, sculptors captured in stone the moment Daphne metamorphoses into a tree in Apollo's arms.

The scientific revolution that happened during the second half of the seventeenth century owes its development to two of the great mathematicians of the age: Isaac Newton and Gottfried Wilhelm Leibniz. The development of calculus by these two great men provided the most staggering shortcut to navigating our dynamic universe. Richard Feynman once described calculus as "the language God talks."

So if you haven't learned calculus yet, now is the time. It will require some equations, but I promise you, they are worth it.

A Universe in Flux

Even before Glenn had completed his orbits of the earth, calculus helped get him up there. As he sat on the launchpad he was aware that the spaceship would need to achieve a particular speed to be able to clear the gravitational pull of the earth, called the escape velocity. But knowing what the speed of the spaceship is at any point as it is propelled into space is not an easy task. Things are constantly changing: the mass of the spacecraft is shrinking as it burns off fuel, the pull of gravity is decreasing as the spaceship travels farther and farther away from the earth, the speed is being affected by the push of the jets and the pull of gravity as they compete. It looks like an impossible cocktail. But the real strength of calculus is that it can accommodate a hugely complex range of changing variables to give a snapshot of what is happening at any particular moment of time.

And it all began with that apple falling from the tree in Newton's garden in Woolsthorpe Manor in Lincolnshire. Newton had retreated back to his family home from his college in Cambridge after plague had hit. Lockdown during a pandemic has certainly been a productive time for some; Shakespeare is said to have completed *King Lear* while the Globe was closed during lockdown. As Newton sat in his garden, he wanted to make sense of the challenge of calculating what the speed of the apple was at any point on its journey from the tree to the ground. Speed is distance traveled divided by the time it takes to travel that distance. That's fine if the speed is constant. But the trouble was that because of the pull of gravity the speed was constantly changing. Any measurements that Newton did would only give him the *average* speed over the period of time he was measuring.

To get a better calculation of the speed, he could take smaller and smaller time intervals. But to get the exact speed at any time really means taking an infinitely small time interval. Ultimately you want to divide distance by zero time. But how do you divide by 0? Newton's calculus made sense of this.

Galileo had already discovered the formula for finding out how far the apple had fallen after any length of time. After t seconds the apple falls a distance of $5t^2$ m. The 5 here is a measure of the particular pull of gravity on earth. An apple tree on the moon will have a smaller number in the equation because there gravity is weaker and the apple falls more slowly. And Glenn's spacecraft would have to keep track of how this number changed as he got farther from the earth.

Let's pick up the apple and throw it directly into the air. I'm going to launch it from my hand at a speed of 25 m/s. Baseball pitchers can hit speeds of over 40 m/s, so this isn't unreasonable. The formula for how high the ball is from my hand after launch then becomes

$$25t - 5t^2$$

I can use this formula to calculate the time it takes to reach my hand again. I need to know when the height above my hand, $25t - 5t^2$, becomes 0 again. Put $t = 5$ into the equation and I get zero. So the total time for the apple to travel up and back down is 5 seconds.

But what Newton wanted to be able to understand is how fast the apple is traveling at each point during its trajectory. The trouble is that this speed is constantly changing as the apple slows down and then speeds up again.

Let's try to calculate the speed after 3 seconds. Speed is distance traveled divided by the time it takes to travel that distance. The distance that the apple travels from 3 to 4 seconds is

$$[25 \times 4 - 5 \times 4^2] - [25 \times 3 - 5 \times 3^2] = 20 - 30 = -10 \text{ meters}$$

The minus sign indicates that it's traveling in the opposite direction to the one I threw it in. It's already heading down. So the average speed over this period is 10 m/s. But that's just the average speed over this one-second interval. It's not the actual speed of the apple at 3 seconds. What if I try to take a smaller time interval? If I keep making the time interval smaller and smaller, what I find is that the speed gets closer and closer to 5 m/s. The instantaneous speed is captured when the time interval becomes zero. This is the snapshot Newton was after. Newton's analysis produced a way to make sense of why the instantaneous speed at 3 seconds should be 5 m/s.

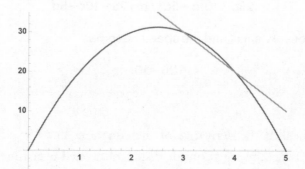

Figure 6.2. The graph of the height of the apple with respect to time. The average speed of the apple between two points in time is the gradient of the line through the graph at these points.

It is possible to interpret this speed in a graph of the distance traveled over time. The average speed between 3 and 4 seconds is the gradient of the line drawn between the two points on the graph at 3 and 4 seconds. As I make the time interval smaller and smaller the line gets closer and closer to the line, just touching the point at time $t = 3$. What Newton's calculus calculates is the gradient of the line that touches at this point, a line called the tangent to the curve. In general, calculus tells us that at time t, the speed and the gradient are given by the formula

$$25 - 10t$$

Here is an explanation. Suppose we are trying to calculate the speed at time t. Let's see how far it travels in a small interval following t, say from time t to time $t + d$.

$$[25(t + d) - 5(t + d)^2] - [25t - 5t^2]$$
$$= 25t + 25d - 5t^2 - 10td - 5d^2 - 25t + 5t^2$$
$$= 25d - 10td - 5d^2$$

Now let's divide by the time interval d:

$$25d - 10td - 5d^2 / d = 25 - 10t - 5d$$

Let d get very small and the speed becomes

$$25 - 10t$$

This is called the derivative of the equation $25t - 5t^2$. This clever algorithm takes the equation for distance traveled over time and produces a new equation that gives us the speed at any point in time. The power of this tool kit is that it doesn't just apply to apples and spaceships. It gives a way of analyzing anything in flux.

If you're a manufacturer, it is important to know what the cost is for creating your product so that you can set a price that will generate a profit. The cost of making the first product is going to be very high because of the cost of setting up the factory, employing workers, and so on. But as you make more and more products the marginal cost for each extra product will change. To start with, the marginal cost will go down because it becomes more and more efficient to make your product. But if you try to push production too far, costs can go up again. Increased production eventually leads to overtime, use of older and less efficient plants, and competition for scarce raw materials. As a result, the cost of additional units increases.

It's a bit like throwing the ball in the air—in the first second the ball travels far, but each subsequent second the ball slows and covers less ground. Calculus can help a manufacturer understand how costs of goods are changing as the output varies and find the sweet spot of how many goods to produce to get the marginal cost to be its smallest.

Newton's shortcut to navigating a world in flux marks the beginning of modern science. I would rank Newton alongside Gauss as one of the all-time great shortcutters. I've even gone as far as making a pilgrimage to Woolsthorpe Manor, where Newton is reputed to have sat under the apple tree that sparked the creation of this inspired shortcut. I was amazed to see that the tree is still there! The person who showed me around allowed me to take two of the apples from the tree, and I managed to grow an apple tree in our garden from one of the seeds. I spend many hours sitting under that tree hoping to find the shortcut that will guide me to the other side of the problem I'm currently working on.

Like me, Gauss was a huge fan of Newton's work. "There have been only three epoch-making mathematicians: Archimedes, Newton, and Eisenstein," he wrote. That last one is not a misprint; Einstein hadn't even been born. Gotthold Eisenstein was a young Prussian number theorist who impressed Gauss by solving a couple of problems that Gauss couldn't crack.

Gauss was always rather skeptical about the story of the apple being key to Newton's discoveries:

The history of the apple is too absurd. Whether the apple fell or not, how can any one believe that such a discovery could in that way be accelerated or retarded? Undoubtedly, the occurrence was something of this sort. There comes to Newton a stupid, importunate man, who asks him how he hit upon his great discovery. When Newton had convinced himself what a noodle he had to do with, and wanted to get rid of the man, he told him that an apple fell on his nose; and this made the matter quite clear to the man, and he went away satisfied.

It's true that Newton had little interest in publicizing his ideas. For Newton, calculus was not so much a device for optimizing solutions as a personal tool that helped him reach the scientific conclusions that he documents in the *Principia Mathematica*, the great treatise that he published in 1687 describing his ideas on gravity and the laws of motion. He wrote that his calculus was key to the scientific discoveries contained inside: "By the help of this new Analysis Mr Newton found out most of the propositions in the Principia."

He liked to refer to himself rather grandly in the third person. But no account of the "new analysis" was published. Instead he privately circulated the ideas among friends, but they were not ideas that he felt any urge to publish for others to appreciate. This decision not to formally publish his ideas was to have ugly consequences, because some years after Newton's discovery another mathematician also came up with the mathematics of calculus: Gottfried Leibniz. And it was his approach that highlighted the optimizing power of this tool.

Maxing Out

While Newton needed calculus to understand the fluctuating physical world that surrounded him, Gottfried Leibniz came at the idea from a more mathematical, philosophical direction. He was fascinated by logic and language and was motivated by the idea of capturing a whole range of different things in a state of flux. Leibniz had big ambitions. He believed in an incredibly rationalist approach to the world. If everything could be reduced to mathematical language where everything was unambiguously expressed, then there was hope of putting an end to human strife: "The only way to rectify our reasonings is to make them as tangible as those of the Mathematicians, so that we can find our error at a glance, and when there are disputes among persons, we can simply say: Let us calculate, without further ado, to see who is right."

Although his dream of a universal problem-solving language was not realized, Leibniz was successful in creating his own language that could solve the problem of capturing things in flux. At the heart of his new theory is an algorithm, something like a computer program or a mechanical set of rules that could be implemented to solve a vast array of unsolved problems. Leibniz was very pleased with his invention: "For what I love most about my calculus is that it gives us the same advantages over the Ancients in the geometry of Archimedes, that Viete and Descartes have given us in the geometry of Euclid or Apollonius in freeing us from having to work with the imagination."

Just in the same ways as Descartes's idea of coordinates had converted geometry into numbers, Leibniz's calculus had provided a new language to master and pin down the world of change.

Although Newton and Leibniz are credited with the great breakthroughs on making calculus into the powerful subject taught today, it was Pierre de Fermat, better known for his Last Theorem, who recognized that calculus can find shortcuts to the optimal solution to a problem.

Fermat was interested in trying to find a way to solve the following sort of challenge. A king has promised his trusty advisor a piece of land by the sea for his good services. The king has given the advisor 10 km of fencing to mark out a rectangular plot of land bordered by the sea. The advisor obviously wants to maximize the area of the land. How should he arrange the fences?

Essentially there is one variable he can vary, which is the length of the side of the rectangle perpendicular to the sea, which I will call x. As that grows, the length along the beach that he can capture gets smaller. What is the balance between the two lengths that will make the area of land it surrounds largest? One's first intuition might be to balance the two and to make a square shape. Making things as symmetrical as possible is often a good strategy for finding the shortcut to a solution. A bubble, for example, chooses the symmetrical sphere as the shape that uses the least surface area to enclose the air inside. But is the symmetry of the square the right answer for our trusty advisor?

There is a very simple formula for the area of land depending on x, the length of the side that is varying. Since the beach length is $10 - 2x$, the area A must be

$$x \times (10 - 2x) = 10x - 2x^2$$

What is the value of x that makes this biggest? One strategy might be to just keep trying out values until we get a feel for the x that seems to be making the area biggest. That's the long way around this problem. Fermat realized that there was an easier way.

The way he found the shortcut was to turn the equation for the area into a picture. Draw a graph where the height about each value x is the area of the land. This produces the graph of the equation $10x - 2x^2$. The shortcut will eventually spare you from drawing the picture, but to find shortcuts sometimes you need to take a detour first. The shape of the graph is a curve that climbs from zero area if $x = 0$ through to a peak that then decreases until you get zero area again if $x = 5$. The key is working out where that peak is. That's where the area is greatest. What is the value of x that produces that peak?

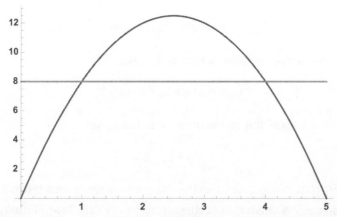

Figure 6.3. The graph of the area of land with respect to the distance along one side. The area is maximal when the horizontal line intersects the graph at one point, not two.

Draw a horizontal line through the graph. In general it will cut the graph at two points, except for that one point right at the top, where the horizontal line will just sit on top of the graph, touching at one point. This is the point we are after—the top of the graph, where the area is biggest. Fermat found a strategy that identifies the point where this happens without having to draw the graph. It revealed that setting $x = 2.5$ would optimize the area of the land. The region was a rectangle, not a square, whose long side was twice the short side. If you are feeling brave enough to do a bit of algebra, here are the details of Fermat's idea.

Suppose I set $x = a$. Then the horizontal line through this point will hit the graph on the other side at a point $x = b$, where the height above this point is the same as the height above $x = a$. So this is the point at which

$$10a - 2a^2 = 10b - 2b^2$$

I can use some algebraic tricks to simplify this. Put the square terms on one side:

$$2a^2 - 2b^2 = 10a - 10b$$

But I can factor the side with the squares:

$$2\,(a-b)\,(a+b) = 10\,(a-b)$$

And now cancel the terms on both sides to get

$$(a+b) = 5$$

But Fermat is interested in the moment when a and b are actually equal, because that is the top of the curve. That is the point where $b = a$. Putting this into the equation, we get

$$2a = 5$$

The point at which the graph tops out is when a is 2.5. This is the length of the side of the rectangle that will maximize the area of land. We actually get a rectangle that is 2.5 by 5.

There is an interesting moment in the calculation above when I divided by $a - b$. That was all OK unless $a = b$, in which case I just divided by 0, which isn't allowed. But hold on. Didn't Fermat actually want to find the place where $a = b$? So does that invalidate the whole shebang?

This is the point of calculus. It makes sense of how to divide by zero.

There were calculations here, but where was the calculus? Calculus gives you the gradient of the line tangent to each point on the curve. Fermat has identified the maximum area as the point when the tangent is horizontal. This is the point when the gradient or the derivative is zero. This is the strategy for using calculus to find optimal solutions to the outputs of equations: find the point at which the derivative of the equation is zero.

The curve describing the area of land looks remarkably like the one Newton drew to keep track of the height of his apple. The equation for the area of the land, $10x - 2x^2$, and the equation for the distance of the apple from my hand, $25t - 5t^2$, are essentially the same equation. The second equation is simply the first one multiplied by 2.5. This is one of the great shortcuts of mathematics: the same equation can cover a multitude of different scenarios. In the case of the apple, this maximum point is the moment that the speed hits zero and the apple begins to move in the opposite direction.

But this sort of equation might represent many other things as well: energy consumption, quantity of building materials, time to one's destination. Having a tool that can find the best way to maximize or minimize these different quantities has been transformative. If the formula is the profit of a company that will depend on various ingredients that your company might vary, who wouldn't want a tool that could tell you what value to set these variable inputs in order to maximize the profit output? Calculus is the shortcut to maximum profitability.

Mathematical Scaffolding

Although calculus was principally created to analyze the world as it changed over time, it was also good at analyzing change outside of time. In particular, calculus has become a very powerful tool for looking at the different ways you might design a building and to find the version that optimizes energy efficiency, acoustic quality, or building costs while still making a structure that stands the test of time.

One such building completed in 1710 still stands proud not too far from where I live in London: St. Paul's Cathedral. I have a soft spot for this building partly because it was designed by a mathematician who trained at the same college in Oxford that I attended as an undergraduate. Before becoming one of England's leading architects, Christopher Wren had cut his teeth learning mathematics at Wadham College. As a student he equipped himself with a whole range of techniques that would allow him to find shortcuts to erecting some of the great buildings across the country.

One of his first great achievements was the Sheldonian Theatre in Oxford, the building where we give our students their degrees. The beauty of this building is the fact it has a huge roof with no columns supporting it. This was apparently not so parents could see their loved ones getting their degrees but because the space was principally used for dancing. The way Wren achieved this extraordinary expanse of roof with no visible supports was to use a lattice structure of beams that lie over and underneath. This moves the load-bearing to the beams at the edge, which sit on the perimeter wall. But in order to find the arrangement that would work, Wren was faced with solving twenty-five simultaneous linear equations. Despite having trained as a mathematician, he was somewhat defeated by the problem and ended up having to get help from the Savilian Professor of Geometry, John Wallis. Asking for help is often an important shortcut!

But it was building the dome of St. Paul's Cathedral where Wren's mathematics would truly come to the fore. The dome that you see from the outside as you approach the cathedral is spherical in shape. A sphere has a beauty and perfection about it that are particularly appealing when seen from a distance. The shape also tapped into the idea of the church representing the shape of the cosmos. But the sphere has a crucial flaw when it comes to buildings: it is not a shape that stands by itself. Indeed, the shape is too shallow to support itself, and the peak of the dome would come tumbling down into the heart of the cathedral. So St. Paul's Cathedral consists not just of one dome but of three different domes.

The dome you see when you are within the cathedral is not actually the inside of the external dome. It is in fact a second dome whose shape is made up from a new curve called a catenary, later explicitly identified by Leibniz (among others) using calculus, that is able to stand freely without any support. This is the shape that a chain makes when it is hung from two ends. Just as a ball left to roll on a mountain finds the point of lowest energy to rest, the chain too minimizes the amount of potential energy it possesses. Nature is very good at finding these low-energy states. But the key point for an architect such as Wren is that this low-energy solution, when turned upside down, becomes the shape that can support its own weight and stand without support.

But what is the shape of this curve? What Leibniz was able to do was to vary the shape of the curve and produce an equation for the potential energy contained in each of those curves. He could then use the calculus to identify the curve with the smallest energy. This would be the shape that the chain would hang in. Once the shape had been identified, it could then be used by subsequent generations of architects to make freestanding domes without having to scale up physical chains hanging in the space. Wren particularly liked a dome in the shape of the catenary because when you look up it creates a forced perspective making it feel higher than it actually is. The use of mathematics to create an optical illusion is a big theme of the architecture of the Baroque period.

There was still the problem of how to make sure the external dome wasn't going to collapse into the cathedral and destroy this beautiful inner dome. This is why there is a third dome that is hidden between the two domes you can see. On a visit to St. Paul's Cathedral recently I got the chance to get inside the two domes and to see the third dome that is doing all the work of supporting the external spherical dome. This third, hidden dome uses the catenary curve again to determine the shape that Wren would need to support the cupola that sits at the top of the outer dome. If you hang a weight on a chain, then it pulls the chain down. Calculus can be used again to give a mathematical description of this new shape with minimum energy. But the clever thing is that if you turn this new shape upside down, then this arch can support a weight sitting on top of the arch equivalent to the weight you've used to pull the chain down. This is how Wren worked out the shape of the inner dome that is supporting the top of the spherical dome that you see from outside.

The most extraordinary use of these weighted chains to build domes can be found if you go down into the basement of the Basilica de la

Sagrada Familia in Barcelona. Antoni Gaudi used the principle for the design of the roof of his unfinished chapel. He attached huge numbers of bags of sand to a web of strings that hung in these catenary curves to represent the load of the structure that needed to be supported. Turn the shape that these strings were making upside down, and you had the shape of a roof that could be built without falling down. By adding and moving the bags around, Gaudi was able to create the shape of the chapel roof that he was after, while still being confident that it wouldn't fall in when he tried to build it. But for producing a mathematical description of all those curves that could then be given to the manufacturer, you are going to need the shortcut of the calculus. Today's architects have replaced chains and bags of sand manipulated by hand with calculus and equations manipulated by computers in order to create the curvaceous buildings that grace our city skylines.

But it isn't just cathedrals and skyscrapers that calculus helps build. Another of Leibniz's successes in finding curves with optimal properties is the discovery of the best curves for making roller coasters!

Roller Coasters

I love roller coasters. It's not just the thrill of the ride. If you're a nerdy mathematician like me, it's the buzz of all the geometry and calculus that has gone into constructing a ride that pushes things to the limit while keeping the train attached to the track. There is one roller coaster in Europe that gets my mathematical blood racing more than any other: the Grand National roller coaster in Blackpool. When you race around this track, you are actually experiencing not just the power of calculus but also one of the most exciting shapes in the mathematician's cabinet of curiosities: the Möbius strip.

As the name of the ride suggests, the Grand National is a race between two trains. When you get in your carriage at the top of the ride there appear to be two parallel tracks. Riders can touch each other as the tracks race through the twists and turns, passing through features named after some of the notorious jumps of the famous horse race. But as the trains make the final dash for the winning post something rather strange happens. The trains arrive at the opposite stations to the ones they embarked from. Very curious. The tracks never meet and cross each other. How on earth did the designers create this feat?

The effect is achieved at the infamous Becher's Brook jump, where one track races over the top of the other. From that point the tracks swap sides, so by the end of the ride the trains arrive at the opposite station.

And it is this simple twist at Becher's Brook that is the key to the Möbius strip, the beautiful mathematical shape that underpins the design of this track. To construct your own Möbius strip, take a long strip of paper about 2 cm in width. Now join the ends up in a loop, but before you join the ends, make a twist of 180 degrees in one end. If you imagine a piece of paper running between the two tracks of the Grand National ride, then at Becher's Brook the paper is twisted 180 degrees as the tracks run under and over each other before they are joined to the track at the beginning of the ride.

The Möbius strip has some very curious properties. The shape has only one edge. Put your finger on the edge of the strip and follow it round. You will be able to reach any other point on the edge. This means that the roller coaster at Blackpool is actually just one continuous track rather than two parallel tracks. But what roller coasters like the one at Blackpool really want is speed!

If you want the fastest roller coaster, then it turns out that calculus will help. In fact, this is the challenge I set at the beginning of the chapter: Given two points A and B in a vertical plane, what is the curve traced out by a point, acted on only by gravity, that starts at A and reaches B in the shortest time? This was a problem first posed not by the creator of a theme park but by Swiss mathematician Johann Bernoulli in 1696. He chose it as a challenge to the two great minds of the day: his friend Leibniz and his adversary in London, Newton:

I, Johann Bernoulli, address the most brilliant mathematicians in the world. Nothing is more attractive to intelligent people than an honest, challenging problem, whose possible solution will bestow fame and remain as a lasting monument. Following the example set by Pascal, Fermat, etc., I hope to gain the gratitude of the whole scientific community by placing before the finest mathematicians of our time a problem which will test their methods and the strength of their intellect. If someone communicates to me the solution of the proposed problem, I shall publicly declare him worthy of praise.

The challenge was to design a ramp that will get the ball from the top point, A, to the bottom point, B, in the fastest time possible. You might think that a straight ramp would be quickest. Or possibly an inverted curved parabola, like the path that a ball follows when it is thrown in the air. In fact, it's neither of these. The fastest path turns out to be a shape called a cycloid—the path traced by a point on the rim of a moving bicycle wheel.

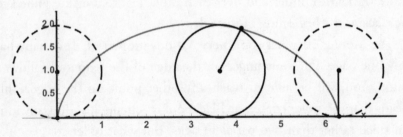

Figure 6.4. The cycloid: the curve traced out by a point on a circle as it rolls along a straight line

If I invert this curve, then this is the fastest way to get from A to B. The curve often descends below the level of the destination, gathering more speed that is then used to get it to its destination up the final climb before any of the other curves deliver the ball.

Because calculus can find the minimum and maximum solutions as the variables range over a particular set of constraints, it doesn't matter that there are infinitely many curves from A to B. The equations will always allow us to find the fastest.

Newton and Leibniz ended up in a terrible fight over who discovered this amazing shortcut for finding the optimal solutions to problems. After years of acrimony and accusation, in 1712 the Royal Society in London was asked to adjudicate between the rival claims: Was Newton's method of fluxions (as it was known) discovered first, and had Leibniz plagiarized his ideas with his invention of the differential method? Two years later the Royal Society officially credited Newton with the first discovery of the calculus and, although recognizing Leibniz as the first to publish, accused Leibniz of plagiarism. However, the Royal Society's report was probably not the most impartial: it was in fact written by its president, one Sir Isaac Newton.

Leibniz was incredibly hurt, because he admired Newton; he never really recovered. The irony is that it is Leibniz's account of calculus that eventually triumphs, not Newton's.

Although the ideas underlying Leibniz's development of the calculus had much in common with those underlying Newton's, there was a big difference. Leibniz was coming at his calculus from a more linguistic, mathematical direction. Leibniz was not concerned with capturing the speed of a falling apple over time; he was considering a much more general setting. If something's behavior depended on several factors, his calculus was designed to study how that behavior varied when you changed the things it depended on.

Newton was really a physicist at heart. He was probably handicapped by his goal of describing the physical world. The language and notation that Leibniz introduced were far more flexible and able to cope with different settings. And it is Leibniz's notation that has stood the test of time. The dy by dx's taught in schools and universities is Leibniz's notation. That integral sign that looks like an elongated S was first drawn in Leibniz's paper of 1686. Newton had a system of tiny dots that he would put over a quantity whose rate of change he was trying to measure. But his dots couldn't distinguish whether the equation was measuring change with respect to time or some other quantity such as position or temperature.

If the truth be told, both Leibniz and Newton had only begun the whole process of developing calculus. Both men's treatises and analysis left a lot to desired. It would be up to the next generation to put calculus on sound logic foundations. But there is no denying that the advances of the next generation were made possible by the breakthroughs that both Newton and Leibniz had made. As Newton once famously said: "If I have seen further it is only by standing on the shoulders of giants."

Do Dogs Do Calculus?

But perhaps Newton and Leibniz were both beaten to the discovery of the calculus by another adversary. There is evidence that the animal kingdom has known how to home in on optimal solutions long before humans came up with the shortcut of the calculus.

Let's return to our trusty advisor, who is now relaxing on the beach after having claimed the maximum bit of land possible using his calculus. He suddenly sees a swimmer out at sea in distress. He shouts to the lifeguard on the beach to rescue the swimmer.

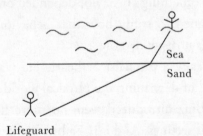

Figure 6.5. What is the quickest path for the lifeguard to reach the drowning swimmer?

Assuming the lifeguard can run twice as fast as she can swim, where should she enter the water to achieve the quickest rescue?

If the lifeguard were trying to minimize the distance traveled, then she would just draw a straight line between the start and finish. But since the lifeguard is slower at sea than on land, she wants to choose a path that cuts down on the time spent at sea. But if she goes for the point that minimizes the time at sea, there is a problem: it means a longer path across the beach, which could end up increasing the total time. The optimal path looks like it will send the lifeguard to the right of the center but not quite as far as the point where the line from the swimmer to the land is perpendicular. But where is the best place to enter the water to find the true shortcut to the drowning swimmer?

This was another of the problems that Fermat thought about. It is again an optimization problem: how to find the fastest path to rescue the swimmer. Fermat encountered the problem not in terms of a lifeguard finding the fastest path but as the challenge of finding the path that a light beam would take.

You might have experienced the rather strange illusion that happens in a swimming pool that a stick inserted into the water appears to suddenly bend as it enters. It is not the stick but the light traveling from the stick to your eye that is bending. As I described in Chapter 4, light loves taking shortcuts. It tries to find the fastest way to travel from the stick to your eye. But light travels more slowly in water than in air. So just like our lifeguard, it tries to spend as little time in the water as possible while balancing that against not making the time in the air too long. The same explanation is key to the strange experience of seeing a mirage in the desert. The light from a patch of sky exploits a shortcut through the warm air closer to the ground before sweeping back up to your eye, making the sky appear to sit in the desert looking like water.

Just as the advisor did with his fence, the lifeguard needs to cook up an equation for the time it takes to reach the swimmer based on entering the sea x meters from the start point. Then, by applying the tool of calculus, she can find the value of x that makes this time smallest. But what if you don't have access to pen and paper? What if you haven't invented algebra and calculus? What if you are just relying on intuition and feeling? What if you were a dog? How good is a dog at judging the right place to enter the water?

Tim Pennings, then a professor of math at Hope College in Michigan and a dog owner, decided to do some experiments to see whether his dog was any good at cracking this calculus problem. Like many dogs, his Welsh corgi, Elvis, was a sucker for chasing a ball. So Pennings decided that instead of the challenge of rescuing a drowning swimmer, he would experiment with chucking a ball into Lake Michigan during the walks he took with Elvis by the lake and see which path Elvis would choose to retrieve the ball.

Of course, it might be possible that Elvis's main objective would be to minimize the amount of energy expended in retrieving the ball, in which case the smart solution was to take the time in the water to its minimum and run to the point where his path in the water was perpendicular to the shore. But Pennings could see by the glint in Elvis's eye and the elevated excitement level as the ball left his hand that retrieving the ball as fast as possible was going to be the goal. The stage was set for his experiment into Elvis's intuitive grasp of calculus.

On a day when the waves on Lake Michigan were low and the ball wouldn't move too much once it landed in the water, he set off with Elvis. With the help of a friend, Pennings launched the ball into the water, then ran after Elvis, slamming a screwdriver into the ground at the point where Elvis entered the water and then, running alongside the water, measuring with a tape measure how far Elvis swam before retrieving the ball.

There were a number of false starts when Elvis just ran directly into the water in a clearly suboptimal route. Pennings decided to delete these data points from his analysis. As he said: "Even an A student can have a bad day." But by the end of the day he'd managed to gather thirty-five data points for Elvis's solution to the challenge. So how did Elvis do? Remarkably well! In most cases Elvis was close enough to the optimal point of entry. The variables that obviously existed in the experiment could easily have accounted for Elvis's approximation.

Does that mean Elvis knows the shortcut of calculus? Of course not. But it is striking how the brain has evolved to find these shortcuts without the power of formal mathematical language. Nature favors those who can optimize solutions, so animals with brains that could intuitively solve these challenges survived more often than those who got it wrong. But there are limits to what an intuitive brain can estimate. That's why, sitting on the launchpad at Cape Canaveral, John Glenn wanted the numbers to be run through this advanced tool we've developed called the calculus to work out the best path home rather than trusting to his intuition.

Sometimes animals would use teamwork to solve the problem Elvis the dog had been challenged with. There is evidence that an ant colony faced with a challenge similar to the lifeguard problem will also do as well as Elvis in finding the optimal path. The ball this time was replaced by food. In an experiment conducted with fire ants by a team of researchers from Germany, France, and China, the ants found optimal paths across two different domains to retrieve food for the nest. Here they have the chance for many ants to strike out and experiment with different routes. The ants lay pheromone trails for other ants to follow. As more ants home in on the optimal solution, that path's pheromone trail gets stronger.

The ants actually do something similar to how we believe light finds the optimal path. How does one photon of light know to find the optimal path? Quantum physics asserts that the photon of light goes into superposition and tries out all the paths simultaneously, collapsing into the optimal path once observed. The ants use a similar strategy of trying everything out using lots of ants before finding the best path.

Nature is very good at finding optimal solutions. Light finds the quickest path to its destination. In modern physics, gravity is interpreted as matter falling through the geometry of space-time finding the quickest journey through that geometry. Hanging chains solved the problem of creating a stable dome for Wren. Bubbles are spherical, taking advantage of the shape's minimal energy use. In more recent times, soap films were used by Frei Otto to design the 1972 Munich Olympic stadium. The strange undulating canopy that covers the stadium is structurally stable thanks to Otto's analysis of how soap bubbles form on the metal frame.

This strange property of Nature homing in on low-energy optimal solutions was captured mathematically in the first half of the eighteenth century by Pierre Louis Maupertuis in his principle of least action. As Maupertuis explained, the mathematics translates into the dogma "Nature is thrifty in all its actions." Quite why Nature is so parsimonious is still something of a mystery. But sometimes dogs or ants or soap films aren't on hand to help us find the answer we're after. Instead we can turn to this incredible tool created by Newton and Leibniz. Calculus has been and will continue to be the most amazing shortcut to the optimal solutions to the challenges we face. As Gauss, the ultimate shortcutter, himself observed about calculus: "Such conceptions unite, as it were, into an organic whole countless problems which otherwise would remain isolated and require for their separate solution more or less application of inventive genius."

SHORTCUT TO THE SHORTCUT

Although calculus is one of our greatest shortcuts, it does take some technical expertise to be able to employ this tool. Even if the idea of doing a crash course in calculus is something most people aren't interested in, it is at least worth knowing that there exists this technique to find optimal solutions. Many shortcuts require a technical guide to help us navigate potentially tricky terrain. So if you've got parameters that can be varied and you want to know the optimal setting for these variables, then contacting an expert in the calculus might be your best shortcut. As Newton recognized, standing on the shoulders of giants has always been a clever shortcut. And sometimes you might find that the technical guide isn't your local neighborhood mathematician but Nature. It is always worth looking to see if Nature has already homed in on an optimal solution to your problem. A soap film might already reveal the low-energy solution to an engineering problem. The path of light might be pointing you in the direction of the shortcut. Or perhaps following a colony of ants will shortcut trying out too many options.

PIT STOP 6

ART

O NE OF THE KEY LESSONS of mathematics is the power of an algorithm to shortcut hard work. Instead of treating each problem on a case-by-case basis, an algorithm crystallizes what unifies all the problems and then presents a recipe that anyone can apply, regardless of their particular setting. Calculus is one such algorithm. It doesn't matter whether your equation is describing profit margins, the speed of a spacecraft, or energy consumption; calculus is an algorithm you can set on each scenario to find the optimal solution.

I was rather surprised to discover that algorithms might also help make art. I learned this in a recent conversation with Hans Ulrich Obrist, curator at the Serpentine Gallery in London. I was intrigued because I've always been terrified by a blank canvas, and I wondered whether having a shortcut might help me turn my creative ideas into something real.

Obrist's idea grew out of the challenges of globalization in the art market. At the start of his career, the art world was still oriented toward the West. An exhibition would travel to Cologne or New York and might take a detour to London or Zurich. But with art galleries opening across the globe, Obrist was eager to crack the challenge of how you get a new exhibition into spaces in South America or Asia. It becomes logistically challenging to take a big exhibition to all the spaces that started wanting to host exhibitions. So Obrist, in collaboration with artists Christian Boltanski and Bertrand Lavier, came up with a way to overcome this: an exhibition entitled *do it*. This involved creating a set of instructions or "recipes" for a piece of art that someone else can use to make it where they are, whether they are in China, Mexico, or Australia.

For Obrist *do it* was the shortcut to the challenge of globalization. You don't bother trying to transport material works in big crates. Just create instructions that can be realized anywhere and in simultaneous time frames. A generative exhibition. An artistic algorithm. The instruction becomes the shortcut. These *do it* instructions are similar to musical scores, which—in, say, the form of an opera or symphony—go through countless realizations as they are carried out and interpreted by others.

The idea of instructional art isn't new. It has its origins in the work of Marcel Duchamp, who in 1919 sent instructions from Argentina for his sister Suzanne and Jean Crotti to make his gift for their marriage. To create the oddly named wedding present, *Unhappy Ready-Made*, the couple was told to hang a geometry text on their balcony so that the wind could "go through the book and choose its own problems." Art by instruction exploded in the late 1960s, driven by the works of John Cage and Yoko Ono. But it was Obrist who realized that instructions could be more than an interesting conceptual idea—they could be a genuine shortcut around the logistical problems of a global art world.

One of the exciting by-products of *do it* is that it has empowered those who might otherwise be frightened of trying to create art. He and I talked during the coronavirus lockdown in Europe, and I learned that Obrist was excited by a new role that the *do it* instructions played during this difficult global period.

"The shortcut became a sponge. Every place it went it could learn and take on board new instructions. So it became this growing archive. We started seeing a Chinese version. A Middle Eastern version. And over the last few months, I've been getting all these messages, first from China and then from Italy and then Spain. Little by little, as all the shutdowns started happening, people started to take their *do it* books from the shelves and realize some of these artists' instructions at home."

As I have experienced in my own work, shortcuts often appear only after a long journey. The same is true for Obrist. "In art we often need detours, and in exhibitions we need detours. But detours are in some ways the opposite of shortcuts. I talked once to David Hockney and he said [that in order to create a painting, first] he needs to write a novel or he needs to do a film or a scientific treatise on perspective or he goes on the iPad and makes iPad drawings. It always leads him back to painting, but it's almost as if he needs these detours."

When Obrist set out on his *do it* project, he began simply with a little booklet with twelve instructions for people to try. But the idea soon began to take on a life of its own as more and more artists responded to the initiative.

"The project seemed so straightforward, but it turned out to be my most complex project ever, with many side routes and detours. It became this sort of learning system. I thought that was really fascinating, because I thought that *do it* was an extreme shortcut, because it was this idea that basically you take a route more direct than the one ordinarily taken. With the instructions you go directly from the artist to the one that realizes them; there is no one in between. You can just do it. One can do it more quickly. It would lead to more immediate results. Yet this project has turned out to be my longest project. It's been going on for twenty-seven years. So in that sense, it's a strange paradox that the shortcut was the biggest detour."

For Obrist, these instructions are a bit like a positive virus. A virus spreads so effectively because at its heart is a short bit of RNA that is a set of instructions for how to replicate itself using the cellular material of the host. Interestingly, one of the shortcuts that a virus uses is the mathematical concept of symmetry. A virus is often put together like a symmetrical die. This has the advantage that the same instruction is used at different regions of the shape, meaning that you don't need custom instructions for different regions.

But symmetry also turns out to be a shortcut that another artist has exploited in creating his work. Conrad Shawcross is a sculptor who loves exploring the interface between art and science. His work has been recognized worldwide, and in 2013 he was elected an academician at the prestigious Royal Academy of the Arts. Shawcross's studio is a short bicycle ride from my home in East London, so I was intrigued to meet up to find out if there are any shortcuts that he has used to become an internationally recognized artist. He told me that he considers shortcuts a way of making ambitious achievements manageable: "You have to be very efficient and very smart about your processes in order to achieve something that otherwise would be impossible. It's about creating templates or jigs or repeat parts that can be put together to create complexity."

Shawcross has often been inspired in his work by rule-based artists. He is an admirer of Carl Andre's work, which takes the brick as the element that will be repeated, and of the work of Monet, who would return to the same lily pads at the same time of day to paint the incremental changes. For Shawcross, the seed for many of his early explorations was an important mathematical shape called the tetrahedron, a triangular-based pyramid.

Part of the appeal of the tetrahedron is the ancient Greek belief that this was indeed one of the building blocks of the universe itself. The Greeks thought that matter was made from four basic elements, earth, wind, fire, and water, and that each element had its own symmetrical shape. The tetrahedron was the shape of fire. Shawcross's first piece to explore the tetrahedron as a building block of art was a structure he was asked to build in 2006 at Sudeley Castle. He made two thousand oak tetrahedrons and then spent two weeks trying to assemble them into a structure. The process was unruly and precarious. "They form these non-tessellating, fiery tendrils that can never join back onto themselves. It was driving me rather than me driving it. On the one hand, it was a bit frustrating, but it was also an awakening, a failure which taught me a lot and was the beginning of many themes for me."

Shawcross needed to find a way to make something that was both beautiful and structurally sound. He eventually found the insight he needed from a mathematician, who pointed out that if you take three tetrahedrons, there's only one way to assemble them. Here was a perfect example of the power of symmetry to provide a shortcut. If you try to find a different way to fuse three tetrahedrons together, then you will discover that you can always transform by a rotation the new proposal into the first configuration. Rather than having 2000 building blocks, Shawcross realized he actually had a larger building block made out of these three fused tetrahedrons.

"It immediately cut my problem down by a third," he recalled. "Suddenly the task became much more surmountable." With this shortcut at hand, Shawcross just needed to find a way to piece together 667 units made up of three tetrahedrons, a task that was much more achievable in the time he had to make the piece.

But when I talked to Shawcross in his studio it turned out that there are some shortcuts he simply won't take. His extraordinary piece called *ADA*, a moving sculpture that maps out complex geometries in space programmed by a series of gears in the machine, made an appearance as part of a dance piece at the Royal Opera House in London. As always, Shawcross was working against a tight deadline, and it was touch-and-go whether the piece would be ready for the evening performance.

As they were painting *ADA*, someone suggested that there was no need to paint the back of the sculpture, as it wouldn't be seen by the audience. A clever shortcut, you might think. But Shawcross just couldn't bring himself to cheat the audience in this way. With all his pieces, even if there are sides that will never be seen, it is important that they be given the same treatment as the parts in view. The audience may not be able to see the back of the piece, but for a sculptor like Shawcross, this was a shortcut too far.

Here are some *do it* artistic algorithms, your shortcut to creating art at home:

Sophia Al Maria (2012)

Locate a television with a generous selection of satellite offerings.
Utilize the Fibonacci sequence of numbers to select channels in
    order: 0, 1, 1, 2, 3, 5, 8, 13, 21, 34, 55, 89, 144, 233, 377, 610, 987,
    and so on.
Alternatively, use a Fibonacci calculator.

Take a photo with a digital device of each channel in passing.

When you have exhausted your satellite channel options as prescribed by the golden ratio, collate the data in the reverse order you have collected it and compile into a mosaic.

The resulting image is a simplistic representation of one edge of the multifaceted media matrix.

Marvel at the stunning mediocrity of our manmade wonder.

Tracey Emin
What Would Tracey Do? (2007)

Take a table. On the table place twenty-seven bottles, all of different sizes and colors. Take a reel of red cotton and wrap it around the bottles, like a strange web that joins them all together. You can, if you wish, take the reel of cotton underneath the table.

Alison Knowles
Homage to Each Red Thing (1996)

Divide the exhibition space floor into squares of any size.

Put one red thing into each square. For example:

- A piece of fruit
- A doll with a red hat
- A shoe

Completely cover the floor in this way.

Yoko Ono
Wish Piece (1996)
y.o. '96

Make a wish.

Write it down on a piece of paper.

Fold it and tie it around a branch of a Wish Tree. Ask your friends to do the same.

Keep wishing.

Until the branches are covered with wishes.

Franz West
Home do it (1989)

Take a broomstick and tightly bandage both the handle and the
bristles with cotton gauze so that the bristles stand on end.

Take 35 decagrams of plaster and mix with the appropriate amount
of water. Distribute the plaster over the entire bandaged
surface. Take another strip of gauze and bandage the plastered
work again. Apply another layer of plaster to totally cover the
work.

Repeat this procedure once again and let the *Passstueck* [adapter]
dry completely.

The result of this procedure is that the object can be used as
a *Passstueck*, either alone, in front of a mirror, or in front of
guests. Deal with it however you feel suitable.

Encourage your guests to act out their intuitive thoughts for
possible uses of the object.

CHAPTER 7

THE DATA SHORTCUT

> Puzzle: You are invited onto a game show. There are
> twenty-one boxes, and inside each box is a cash
> prize. You are allowed to open one box at a time.
> You can keep the money in the last box you opened.
> But once a new box is opened you can't go back and
> claim the previous box's money. The trouble is, you
> have no idea what the size of the prizes might be.
> There could be a box with a million dollars in it. Or
> they could all just contain prizes of less than a
> dollar. Your challenge is: How many boxes should
> you open to give yourself the best chance of getting
> the biggest prize that is in all the boxes?

E VERY DAY WE ARE generating more and more data as we meander
around the expanding digital world that we are helping to popu-
late. Humankind currently produces in two days the same amount of
data it took us from the dawn of civilization until 2003 to generate.
That is a vast digital landscape to explore. Hidden inside that data is
treasure for any company that can discern patterns that might help
predict your next digital move. It is not easy for a data analyst to navigate this data jungle, but mathematicians have discovered a clever suite
of shortcuts that can uncover this treasure without having to survey the
whole territory.

As soon as the scientific revolution ignited in the seventeenth century we began to be overwhelmed by the data we were generating. John Graunt, one of the first demographers, complained in 1663 about "the overwhelming amounts of information" he was inundated with as a result of his studies of the bubonic plague ravaging Europe at the time. But you need those numbers in order to deal with a pandemic. This is why World Health Organization director general Tedros Adhanom Ghebreyesus told a news conference in Geneva that the key to surviving the 2020 coronavirus outbreak was "testing, testing, testing." Without the data, governments would have no idea what resources to deploy, or to where. But without ways to find the signal in the noise, data is useless. The US Census Bureau complained in 1880 that the data it had collected was so extensive that it would take more than ten years to analyze the numbers, by which time it would be inundated by yet more data from the next census in 1890. Tools were needed to shortcut the messages inside these vast swaths of numbers we were generating and collecting.

My hero Gauss was always a fan of data. He reveled in a book full of numbers he was given for his fifteenth birthday, including tables of logarithms and a list of prime numbers at the back of the book. "You have no idea how much poetry there is in a table of logarithms," he wrote. He spent hours trying to tease out some pattern hidden inside the random-looking primes, eventually realizing that there was a connection with the logarithms at the front of the book, a revelation that would lead to the prime number theorem, predicting the chance that a number taken at random is prime.

He'd successfully extended the observations gathered by astronomers of the passage of Ceres through the night sky before it disappeared behind the sun. He signed up to analyze census data from the Hanoverian government, declaring: "I hope to get the editing of the census, the birth and death lists in local districts, not as a job, but for my pleasure and satisfaction." He even spent time analyzing the university's pension scheme for the widows of professors, concluding that, contrary to what others had feared, the fund was in great shape and could afford to pay the widows more.

His success at recovering Ceres from the noise of the night sky was due to a strategy he developed called the method of least squares. If you have some noisy data and you want to plot the most likely line or curve through the data, then Gauss showed that you wanted to make the sum of the squares of the distance to the curve as small as possible.

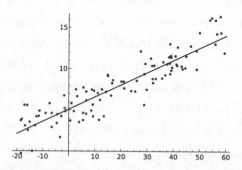

Figure 7.1. Gauss's method of least squares

In the paper he published in 1809 where he outlined the method, he also explains how data likes to distribute itself into a shape we now call the Gaussian distribution. Essentially if you plot many different data sets—people's heights, blood pressure, exam results, errors in astronomical or surveying measurements—then they tend to be spread out, with the majority of cases in the middle and a few outliers at the edges. The curve is often called a bell curve because its shape looks like a bell.

The statistical tools that Gauss and others cooked up are now the go-to shortcuts for anyone navigating our modern data-rich world.

Eight out of Ten Cats

As a kid I was always intrigued by a cat food commercial that would often appear on TV. It claimed that eight of ten cats preferred Whiskas cat food, the brand being advertised. What I found curious about this at the time was that I couldn't remember anyone coming and asking our cat about what cat food she preferred. How many cats, I wondered, had they asked in order to be able to make such a bold claim?

You might think that to be able to legitimately make such a claim would involve a huge amount of work. After all, there are estimated to be 7 million cat owners in the United Kingdom. Clearly the manufacturers of Whiskas cat food didn't knock on 7 million doors to check their claim. It turns out that the mathematics of statistics provides an amazing shortcut to finding out the nation's favorite cat food. The point is that in exchange for a certain amount of uncertainty, the number of cats you need to ask turns out to be remarkably small. Suppose that I am happy to tolerate a 5 percent error in the proportion of cats that claim to like Whiskas. I could skip asking 5 percent of the cats in exchange for this added error. That's fine, but 5 percent of 7 million is still only 350,000 cats. If I skip those cats, that still leaves a lot of cats I need to ask.

The thing is that I've got to be really unlucky to have left out 350,000 cats and all of them don't like Whiskas. Most of the time that 350,000 will be split in a fairly similar way to the whole population. So here is the clever shortcut. What if I say that I am happy to take a sample size such that 19 times out of 20 the proportion I get in the survey is 5 percent away from the proportion I would get if I surveyed the whole population of cats? How big would that sample size need to be? What is amazing is that you only need to ask 246 cats to have this level of certainty that you are actually representing the preferences of the entire 7 million cats in the United Kingdom. That is a surprisingly small number of cats. This is the power of mathematical statistics—to be able to provide you with the confidence to make such a claim based on only asking 246 cats. Once I'd taken a course on mathematical statistics I understood why our cat had never been asked what cat food she liked.

Even the ancient Greeks recognized the power of inferring a lot from a little. When an alliance of city-states planned an attack on the city of Platea in 479 BCE, they needed to calculate the size of the ladders they would require in order to scale the city walls. Soldiers were sent out to measure the size of a sample of the bricks used to build the wall. By taking the average size and multiplying it by the number of bricks that could be seen in the wall, they got a good estimate of how high the walls were.

But it wasn't until the seventeenth century that more sophisticated approaches began to emerge. In 1662 John Graunt used data about the number of funerals held in London to give the first estimate of the population of the city. Based on his data gathered from parish records, he estimated that 3 people died each year per 11 families and that the average family size was 8. Given that the number of funerals recorded each year was 13,000, this led him to estimate the population of London to be 384,000. In 1802 Pierre-Simon Laplace went even further and used a sampling of registered baptisms in thirty parishes to reach an estimate for the population of the whole of France. His data analysis of the parishes indicated that there was one baptism for every 28.35 people living in the parish. Given a record of how many baptisms in total had been recorded in France that year, he was able to come to an estimate for the population of France of 28.3 million people.

Even knowing how many cats there are in the United Kingdom requires the kind of statistical shortcut of going from the small to the big. In the case of the United Kingdom's cat population, we can apply a strategy similar to the one used by the soldiers in Greece: survey a small sample and scale up. If you know the proportion of cats per capita in the smaller sample, you can simply multiply by the total population of the country to get an estimate. But what if you wanted to estimate the total number of badgers in the wild in the United Kingdom? We don't have badgers owned by people, so we can't simply use the number of people the way we did for the cats.

Instead ecologists use a clever shortcut called capture-recapture, a strategy that is at the heart of how Laplace made his estimate. Suppose they are trying to estimate the badger population in the county of Gloucestershire. Ecologists will begin by setting up a number of traps to capture badgers over a particular period. But how do they know what proportion of badgers they've caught? They don't. Here's the clever trick, though. Tag all the badgers you've caught and release them back into the wild. Allow the tagged badgers time to reintegrate themselves back into the total population. Then set up cameras throughout the countryside to record badgers. Now you get two different numbers: the total number of badgers sighted, and the number of those that are tagged. This gives ecologists the proportion of tagged badgers among the badgers sighted. They now can scale this up. Given that they know the total number of tagged badgers in the county, and they know now what proportion this represents of the total badger population, they can get an estimate for the total number of badgers in the county.

For example, suppose 1 in 10 badgers in the second sample were tagged and the total number of tagged badgers in the first capture was 100. We can estimate the total badger population as 1000 in order to get the same proportion as our video sightings recorded. In Laplace's case the babies born (whose number is known) in the total population (whose number is unknown) represents the tagged sample, and then counting the number of babies in the thirty parishes (both numbers known) represents the recapture part of the experiment.

This tactic has been used to make estimates of everything from the number of people enslaved in the United Kingdom today to the number of tanks that the Germans were manufacturing during World War II.

The trouble sometimes with shortcuts is that they aren't always pathways to knowledge. Sometimes they can lead you astray, giving you the illusion that you've reached an answer even though the destination the shortcut takes you to is miles from where you want to be. This is one of the dangers of statistical shortcuts. They can cut corners rather than offer a genuine shortcut.

Although you can get away with asking 246 cats to get some insight into the preferences of the population of 7 million cats, you certainly wouldn't hope to get much understanding from a sample of 10 cats. And yet the scientific literature has numerous examples of apparent discoveries based on ridiculously small samples. This crops up quite often in many psychophysical and neurophysiological studies reported in major journals because it's just too difficult to include many people in such research. But can you really infer anything from research done on two rhesus monkeys or four rats?

Unfortunately, headlines trumpet that "8 out of 10 Xs prefer Y" without offering any inkling of what sample size was used, leaving you with little appreciation of how likely the discovery is to be true.

The gold standard when it comes to legitimately reporting a significant discovery is given by the parameters that I set for ascertaining a good sample size for the cat food survey. There I was happy with a sample size that 19 times out of 20 would correctly represent the food preferences of the cat population.

When it comes to scientific discoveries and their potential significance—say, taking a drug to address a condition—the outcome is regarded as significant if there is less than a 1-in-20 chance that the condition would have been cured without taking the drug. Suppose, for example, that you had come up with a spell for making a coin always land heads. Most people would be rather skeptical, so what would you need to do to convince them? You cast your spell and then start to toss the coin. Suppose that you got 15 heads in the toss of 20 coins. Is that an indication that you might be on to something? Maybe. If you calculate the chances of this happening for a fair coin, the chances are smaller than 1 in 20. So the fact that with your spell you got 15 heads means that you would be justified in thinking your spell might actually be working.

Since the 1920s this has been the threshold that you had to pass for your discovery to be considered "statistically significant," which would be acceptable for publication. It's called having a P-value of less than 0.05—a 1-in-20 (5 percent) chance that the thing might happen randomly.

The trouble is that if you have twenty research groups, one is very likely to get this result. Nineteen groups would have moved on to other ideas, but the twentieth group would have gotten extremely excited and believed they'd passed the threshold for publication of a significant result. You can see why, with such a threshold, so many crazy hypotheses might have made it into the literature. Conversely, a result that has a P-value of 0.06 (it has a 6 percent chance of happening under the null hypothesis) is regarded as too weak to be statistically significant, and so it will often be rejected. Yet that is equally dangerous as a reason to reject a hypothesis. Negative results don't make good news stories, so nineteen research groups don't publish their discovery of no connection. This is why there has been a call to try to reproduce many results in the literature that have been published because they passed this test of statistical significance.

Great care needs to be taken with the thresholds you use. If you are trying to ascertain whether a coin is fair or not, then this threshold might be OK. But imagine that you are trying to determine whether a doctor's rate of negative outcomes is due to malpractice. You wouldn't want to be bringing in 1 in 20 doctors for investigation. And yet at what point should you start to be concerned?

For example, in September 1998, Dr. Harold Shipman, a well-respected family doctor, was arrested for injecting at least 215 of his patients with lethal doses of opiates. But a team of statisticians led by David Spiegelhalter believes that by using a test that was originally introduced during World War II for quality control of military supplies, they could have detected something strange in Shipman's data much earlier and potentially saved 175 lives.

Thresholds for significance need to be handled carefully. In March 2019, 850 scientists wrote a letter to the journal *Nature* striking back at what they regarded as the crazy obsession that the scientific community has had with the P-value as the benchmark for scientific discovery. "We are not calling for a ban on P values," they wrote. "Nor are we saying they cannot be used as a decision criterion in certain specialized applications (such as determining whether a manufacturing process meets some quality-control standard). And we are also not advocating for an anything-goes situation, in which weak evidence suddenly becomes credible. . . . [W]e are calling for a stop to the use of P values in the conventional, dichotomous way—to decide whether a result refutes or supports a scientific hypothesis."

Wisdom of the Crowd

One clever shortcut that statistician Sir Francis Galton came up with is to consult lots of ordinary people, getting them to do all the hard work, and then use a clever bit of math to finish the job. Galton is rightly criticized today for his immoral racist theories of eugenics, but his theory of the wisdom of the crowd is still considered a valuable tool in analyzing big data. He actually stumbled on his discovery when he was trying to prove the opposite. Indeed, he had such little faith in the collective wisdom of society that he was very critical of allowing the public a say in politics, "the stupidity and wrong-headedness of many men and women being so great as to be scarcely credible."

Hoping to prove his point, Galton decided to do an experiment. The county fair in his hometown, Plymouth, had a competition to guess the weight of an ox after it had been slaughtered and dressed. The challenge attracted eight hundred people, who for sixpence submitted an estimate. Although a few entrants might have been farmers with a bit of experience, most were visitors with little knowledge to tap into. "The average competitor was probably as well fitted for making a just estimate of the dressed weight of the ox, as an average voter is of judging the merits of most political issues on which he votes," Galton wrote dismissively.

But when he analyzed the guesses, he got something of a shock. Although many of the guesses were way off the mark, some grossly underestimating the weight, others greatly overestimating, he discovered that if you took the average of all the values, it was amazingly close to the true value. (Galton actually started by using the value exactly sitting at the halfway point, called the median, which also turned out to be very accurate.) The average of the crowd's guesses came to 1,197 pounds, while the true value was one pound off, at 1,198 pounds.

Galton was shocked: "The result seems more credible to the trust-worthiness of a democratic judgement than might have been expected." His discovery gave rise to a phenomenon now known as the *wisdom of the crowd*: a large set of guesses is, in the aggregate, likely to be correct. To find the aggregate only requires a bit of math, and then you've shortcutted your way to a solution.

I recently received a thank-you note from a member of the public who had used exactly this strategy at his local fair after hearing me speak once about the idea. The challenge was estimating the number of jelly beans in a jar. He'd waited until the final moments of the fair before loading all of his fellow fairgoers' guesses into an Excel spreadsheet, taking the average, and using that number for his guess. It turned out that his guess using the wisdom of the crowd was closest—just 5 beans away from the actual number of 4532. He included a few jelly beans with his letter as my cut of the prize as a reward for informing him of this shortcut.

Another example of the wisdom of the crowd crops up in the famous game show *Who Wants to Be a Millionaire?* Most of the time you're on your own answering questions in your attempt to get fifteen correct and bag the top prize. However, there are a couple of lifelines that you can cling to if you haven't got a clue. One of them allows you to phone a friend, while a second option is to ask the audience. A team of academics from Switzerland gathered data on the German version of the show, which revealed that of 1337 times when the audience was asked, they got it wrong only 147 times. That is a remarkable hit rate of 89 percent. Compare that to the statistic for phoning a friend, which failed to get the right answer 46 percent of the time.

If you are going to use the audience, it is important not to give your views on the possible answer. As a species, we are terribly susceptible to being led astray. Take the case of the contestant who had to answer the following question:

The Norwegian explorer Roald Amundsen reached the South Pole on 14th December of which year?

A: 1891

B: 1901

C: 1911

D: 1921

She was pretty convinced that Robert Falcon Scott, whom Amundsen had beaten to the pole, was Victorian, so she was sure that answers C and D were incorrect. But she really didn't know which of the first two might be right. So she asked the audience. Look at the results she got back.

A: 28 percent

B: 48 percent

C: 24 percent

D: 0 percent

Of course, one's instinct is to go for B. But look at answer C. Why have so many people gone for that answer given that the contestant was pretty sure that it couldn't be correct? The answer is that the contestant was wrong. Indeed, she probably led many people astray by airing her thoughts, ending up with people who voted for B when if left to their own devices they would have voted for C, the correct answer.

However, the tactic of trusting the audience might depend on which country you are playing in. Apparently Russian audiences are notorious at leading contestants astray, deliberately choosing the wrong answer. Of course, you could always try the shortcut that Major Charles Ingram is accused of using to win the top prize: cheating. He apparently had someone in the audience who would cough whenever the host read the correct answer. It turned out, though, that if he'd known his math, he'd have been OK without a coughing assistant. The last question on which the grand prize depended was to identify the name of the number consisting of 1 followed by 100 zeros. Is it (A) a googol, (B) a megatron, (C) a gigabit, or (D) a nanomole? If you need any help, I'd be coughing for answer (A).

If the crowd is so wise, who needs experts? Well, it all depends on the task at hand. Despite the Conservative politician Michael Gove's declaration during the Brexit debacle that "we've had enough of experts," I wouldn't want to get on a plane flown collectively by the passengers. And it doesn't matter if you amassed every amateur chess player in the world to play collectively against Magnus Carlsen—I still know whom I'd put my money on. On which questions might the crowd provide a shortcut to the answer, and on which will the crowd lead you astray? One of the key pointers is to make sure that everyone in your crowd is answering independently. Remember how the player in *Who Wants to Be a Millionaire?* influenced the audience by her belief that the explorer Scott was Victorian?

Solomon Asch illustrated how the crowd can be especially persuasive in influencing people to go against their instincts. In an experiment conducted in the 1950s Asch asked a group of seven people to identify which one of three lines had the same length as the line on the left.

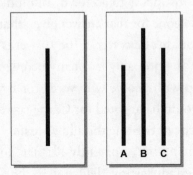

Figure 7.2. Asch's experiment: Which line has the same length as the line on the left?

The twist was that the first six people to answer were in league with the experimenter. They were all told to nominate B as the answer. Time and again, when it was the seventh person's turn to respond, they would not trust their eyes and answer C. Instead the desire to conform with the choices of the group overrode what they were seeing, and they would respond with the same choice as the first six participants.

In an age of social media, this desire for conformity is potentially having a devastating effect on our ability to choose independently of others. Social media makes it quite hard to keep the crowd independent.

But there is some evidence to suggest that total independence isn't necessarily great for making a wise crowd either. In a fascinating study conducted by a team based in Argentina it was discovered that if you allow some conferring among members of the crowd before aggregating the results, the answer beats the response given by a totally independent crowd.

In a live event in Buenos Aires the research team began by asking the 5180 audience members to individually answer eight questions without talking to their neighbor. For example, "What is the height of the Eiffel Tower?" Or "How many goals were scored in the 2010 World Cup?" The answers were then collected and the average answer was calculated. But then the researchers asked the audience to divide into teams of five and to discuss the questions before again being asked for their revised answers. When these were collected and aggregated, the results were much more accurate.

The point is that a few people will have some expert knowledge that might help steer those who frankly haven't a clue. And so the crowd can benefit from some expertise. For example, if you haven't a clue about soccer, then your estimate of the number of goals scored in the World Cup is going to be a total guess. But if in your team of five you get someone who knows a bit about soccer and explains that on average you might get 2 or 3 goals per game and also that there are 64 matches played in the World Cup, then you've got good material on which to base a guess. That would give you, say, $2.5 \times 64 = 160$ goals. The actual answer was 145. But the point is that when you go back to make your new guess based on the discussion you've had, you could well take into account what sounded like the quite convincing expert knowledge offered by one member of your group.

Of course, there are people who confidently believe they are experts and so can lead people astray, so we don't want the crowd influenced by just a single confident leader. But it seems like this combination of a crowd of small teams is more effective than a crowd of individuals.

Another quality that can make a big difference is making sure the crowd has a diverse range of opinions. The audience that took part in the Buenos Aires event might come from a particular social class that is especially interested in attending such events, and as a result you might miss out on a more diverse spectrum of society. This has been illustrated in some interesting cases where the public was asked to help in making budgetary decisions rather than leaving them up to the politicians (a process known as participatory budgeting). The idea was first studied in Porto Alegre in Brazil in 1989. When Iceland's economy failed after the financial crash of 2008, the government decided to invite members of the public to help set budgets. The trouble was that the initiative wasn't generally regarded as a success. By inviting people to apply to take part, it meant that only people interested in politics came forward. The group that formed had an inbuilt bias and did not represent the diversity that the system hopes to exploit.

So instead when the same experiment was conducted in British Columbia, people were selected randomly and sent letters with the expectation that they would attend, rather like jury duty. Because people were randomly chosen rather than allowed to self-select, the resulting group had a much more diverse range of opinions and was far more successful at implementing the ideals of participatory budgeting.

Who Wants to Be a Scientist?

The idea of using a crowd as a shortcut to scientific discovery is the key to the surge of citizen science projects that we have witnessed over the last few years. One of the most successful has been run out of my university in Oxford. The astronomy department is interested in classifying the different types of galaxies that our universe contains. The only trouble is that although we have a fantastic range of telescopes taking wonderful pictures of the galaxies, there just aren't enough research students in the department to look through all the images. When the project kicked off, computer vision was still in its infancy and was unable to distinguish between a spiral galaxy and a spherical galaxy.

But for a human, distinguishing between the two was straightforward. Indeed, the team in Oxford realized that you didn't need a PhD in astrophysics to help out. You just needed a large number of eyes looking through the data. So they launched Galaxy Zoo, one of the first modern-day citizen science projects. To take part in the project, members of the public would be given a quick online tutorial to explain what they were looking for. They were shown the difference between a spiral galaxy and a spherical galaxy and then were let loose on the mass of unclassified images that had been captured by telescopes around the world.

By using the crowd, the astronomy department was able to shortcut the huge task of classifying all this data. It's a bit like the moment Tom Sawyer gets his friends to whitewash the fence he has been told to paint as a punishment. He turns work into play and suddenly all his friends are lining up to help paint.

But the Galaxy Zoo crowd went one better. They discovered a completely new sort of galaxy hiding in the data. Some of the images didn't fit into any of the categories that they were asked to label the data with. Professional astronomers had encountered these images and had just dismissed them as anomalies. But the Galaxy Zoo crowd started to run into more and more of these images, which looked like green peas sitting in the blackness of space. A thread labeled "Give Peas a Chance" emerged on the Galaxy Zoo blog asking not to dismiss these green blobs. The humorous wordplay on the John Lennon song led to the galaxies becoming known as "green pea galaxies." The discovery made by the citizen sciences eventually led to a paper, "Galaxy Zoo Green Peas: Discovery of a Class of Compact Extremely Star-Forming Galaxies," published in the *Monthly Notices of the Royal Astronomical Society*.

The idea of using the crowd as a shortcut to scientific discovery is not new. Edmond Halley enlisted the help of two hundred volunteers in 1715 to work out how fast the shadow of the moon swept across the country during the eclipse that happened on May 3. Stationed at various points across the country, members of the public were asked to record the time and duration of the total eclipse. In Oxford, unfortunately, the skies were overcast and so volunteers were unable to contribute any data. The team stationed in Cambridge was luckier with the weather; the only trouble was that they got distracted and missed it! As the Reverend Cotes, who was in charge of the Cambridge team, wrote to Halley, "We had the misfortune to be oppressed by too much company." They were serving cups of tea to the visiting crowd, and by the time they were ready to do their observations the eclipse had finished.

Halley did manage to collect enough data to be able to estimate that the shadow swept over the earth at an impressive 2,800 km per hour. He published the results in the journals of the Royal Society, where he was a fellow.

Emboldened by Halley's success, another fellow of the society, Benjamin Robins, enlisted the help of citizens in an experiment to discover how high fireworks reached into the sky. He had the perfect occasion to conduct his experiment on the night of April 27, 1749, when to mark the end of the Austrian War of Succession King George II held a celebratory fireworks display, accompanied by music composed especially for the occasion by the king's favorite composer, George Frideric Handel.

Taking out an advertisement in *The Gentleman's Magazine*, Robins asked people to record the height of the fireworks from their location:

> For if such as are curious and are from 15 to 50 miles distant from London, would carefully look out in all proper situations on the night when these fireworks are play'd off, we should then know the greatest distance to which rockets can possibly be seen; which if both the situation of the observer, and the evening be favourable, will not, I conceive, be less than 40 miles. And if ingenious gentlemen who are within 1, 2 or 3 miles of the fireworks, would observe, as nicely as they can, the angle that the generality of the rockets shall make to the

horizon, at their greatest height, this will determine the perpendicular ascent of those rockets to sufficient exactness.

This was not some idle research project. Given the importance of rockets to the military, knowing the range of a firework could be very useful in developing weapons. Unfortunately, the instructions that Robins provided in *The Gentleman's Magazine* were so obscure that everybody was put off from taking part except for one gentleman 180 miles away from London in Carmarthen, in Wales. Patiently waiting on a hilltop, he claimed to have seen two flashes at an elevation of 15 degrees above the horizon. Given the curvature of the earth and the Brecon Beacons in the way, it is highly unlikely that he actually saw any of the six thousand fireworks that were set off. Having heard how many fireworks were used in the display and what little impact it had out in Wales, the volunteer thought the whole thing was a huge waste of public money.

The power of the crowd to help in scientific investigation today has been much more successful than Robins's failed attempt. From counting penguins in video footage from Antarctica to folding proteins to discover the key to degenerative diseases, enlisting the crowd has been a very productive shortcut to new insights.

The power of the crowd as a shortcut to knowledge hasn't been missed by the corporate world, either. In fact, the success of Facebook and Google has depended on the crowd freely giving away valuable data in exchange for their services.

Machine Learning

Galaxy Zoo kicked off in 2007, when computer vision was still very poor. However, the last few years have seen a massive step change in the ability of computers to detect what is in an image. This is due to a new way that code is being written, called machine learning, where code changes and mutates through its interaction with data. The power of allowing code to learn from the bottom up rather than trying to work out the code in a top-down manner has provided an amazing shortcut to writing powerful algorithms. The code itself may not be very efficient and slick, but with today's computing power that isn't as much of a problem as it used to be.

One of the great successes of machine learning has been computer vision. The key to this revolution has been the power of statistical analysis. The computer isn't infallible, but that's fine. Provided it gets the right answer most of the time, that is good enough. And this is the point about our original 8-out-of-10-cats shortcut. To get a success rate of 99 percent in distinguishing a cat from a dog requires being exposed to data, but how much data? We don't want to have to give the computer all the pictures of cats and dogs online, of which there are many!

The general rule for training an algorithm to distinguish different categories of images is that you need 1000 images to represent each category that you are hoping the algorithm will be able to identify. To create a cat recognition algorithm, you need 1000 images of cats for the code to learn from. It turns out that for standard machine learning algorithms, more data doesn't really improve the success rate. The algorithm seems to plateau. But for more sophisticated deep learning models, more data does produce a logarithmic improvement.

Knowing how much data you can get away with is essential when it comes, for example, to wanting to know what variables might be impacting sales. Perhaps you think the day of the week has an impact, or the weather, or whether the news is positive. The way to try to understand what is influencing your sales is to gather data. Take the variables that you think might be affecting the sales and then record the sales for different values of all the variables.

To know the minimum amount of data you can get away with to make an informed inference, we can look to regression analysis and the rule of 1 in 10. If there are 5 variables that you are tracking, then $10 \times 5 = 50$ pieces of data would be the ballpark number to glean what the impact of variations of these parameters might have on sales.

But we have to be careful with this kind of shortcut because it can lead us astray. Just as it is important to have diversity in a crowd if you are hoping to glean some wisdom, it's equally true that you need to make sure the data is diverse. When Amazon hoped to develop an AI to help sift through applications for jobs, they used the profiles of current employees as their model. This seemed like a sensible decision given that the company was happy with the quality of employees on the books to date. But when the AI started rejecting any resume that wasn't from a white twenty-year-old male, the company realized that the algorithm was unfairly discriminating against a whole swath of individuals who were applying for jobs. The Algorithmic Justice League, started by Joy Buolamwini, is calling out algorithmic shortcuts that fail to get us to new destinations and instead just bring us back to old prejudices.

It is also important not to track too many variables at any one time, because the more variables you track, the more likely you are to find patterns inside. The dangers of tracking too many variables were illustrated when an fMRI scanner was used in an experiment to examine 8064 regions of the brain to see which might be involved when the participant was shown various images of human expressions. Sure enough, 16 regions showed a statistically significant response. The trouble was that the subject being scanned was a large dead Atlantic salmon. The researchers had been using inanimate objects such as the salmon in order to correct for false positives. But it illustrates the dangers of simply measuring too many things and hoping to pick out a pattern. The team received one of that year's Ig Nobel Prizes, awarded for achievements that "first make people laugh, and then make them think."

As Craig Bennett, who was one of the researchers on the team, explained:

If you have a 1 percent chance of hitting a bull's-eye when playing darts and you throw one dart, then you have a 1 percent chance of hitting the target. If you have 30,000 darts, then, well, let's just say that you are probably going to hit the target a few times. The more chances you have to find a result, the more likely you are to find one, even by chance.

How Much Data Before You Make Up Your Mind?

The game show I described at the beginning of the chapter is actually a good model of many of the challenges we face in life. Your first boyfriend or girlfriend might be amazing, but do you marry them or is there a nagging feeling that maybe you could do better? Perhaps one of the many other people out there is "the one." The trouble is that if you dump your current partner, there's generally no way back. At what point should you just cut your losses and settle for what you've got?

Apartment hunting is another classic example. How many times do you see a fantastic apartment on your first visit but then feel you need to see more before you commit yourself, only to risk losing the first apartment? It turns out that the key to optimizing your chances of getting the best prize possible is mathematics' second-most-popular number: $e = 2.71828 \ldots$ Like π, mathematics' number one number, e has an infinite decimal expansion that never repeats itself. And e is a number that keeps cropping up in all sorts of different settings. For example, it is intimately related to the way the interest accumulates in your bank account.

But e also turns out to be the shortcut to get the best chance of choosing the winning box in our hypothetical game show. The mathematics proves that if you have N boxes, then you need to gather data from N/e of the boxes to get some idea of what the prize money looks like. We know that $1/e = 0.37 \ldots$ This represents 37 percent of the boxes. Once you have opened 37 percent of the boxes, the strategy is to choose the next box that beats all the boxes that you have opened already. This doesn't guarantee you the best prize, but 1 in 3 times you will end up with the maximum amount possible. If you based your decision on seeing fewer or more boxes, then these chances decrease—37 percent is the optimal amount of data to gather before taking the plunge, whether it's boxes in a game show, apartments, restaurants, or even your partner for life. (Though perhaps it's best not to let your partner know you were so calculating when it came to love.)

SHORTCUT TO THE SHORTCUT

Making a decision on which direction to take your ideas for a new project is often improved by doing a survey of people's preferences. As the saying goes, data is the new oil—but it is still important to know how much you need to power your ideas. Too much data and you can drown. Too little and your project won't get off the ground. The shortcut of statistics reveals that you can often get a long way on a surprisingly small sample size. Finding clever shortcuts to collecting data is important as well. As Mark Twain illustrated, it takes one person a long time to paint a fence, but many hands can quickly finish the job. Tapping into the wisdom of the crowd is a way to glean insights, whether it involves setting up a Twitter poll, devising an online game that yields data, or exploiting Google analytics to understand engagement with your website.

PSYCHOTHERAPY

WHEN I FIRST TOLD SHANI, my wife, that I was writing a book about shortcuts, she was horrified. She is a psychologist and believes that often there is no replacement for the deep, protracted work one needs to undergo in psychotherapy in order to rewire the brain. And yet Shani did admit that even therapy has found shortcuts to tackle the huge mental health problems that face society.

Traditionally, the idea of going to a therapist conjures up the image of years spent lying on a couch talking about your childhood. But it turns out that for some conditions there are very powerful techniques that can shortcut these years in therapy. Shani suggested that I talk to Dr. Fiona Kennedy, who, having practiced as a psychologist for many years, now trains others in the range of intensive therapies that have emerged to tackle mental health issues. These interventions can help patients with phobias, anxiety, depression, and post-traumatic stress without the need for years of treatment.

For Kennedy, one of the reasons these therapies have been successful is that they take a more scientific approach. "If you were going to a surgeon to have a heart operation and you had two candidates to do the surgery, and the first said, 'This is my history of heart operations—these are the techniques I've used and these are the success rates I've had,' and the other one said, 'Well, I don't really collect any data, but I am a very creative person that people find totally inspiring. I've done many operations and I've enjoyed them very much,' who would you get to do your surgery?" Although evidence-based scientific thinking hit the world of psychotherapy only recently, it has been key to the successful introduction of these methods into health services worldwide.

Probably the most well-known psychological shortcut is CBT, or cognitive-behavioral therapy. Developed by Aaron Beck in the late 1960s and early 1970s, CBT focuses on how your thoughts, beliefs, and attitudes affect your feelings and behavior, and teaches you coping skills for dealing with different problems.

Kennedy recalls how as a student she took part in an experiment where rats and students were asked to perform various tasks. "The rats beat the students hands down," she notes. "We were all thinking too hard about what was going on." The experiment illustrated how cognition can interfere with a journey to a successful outcome. For Beck and others, the key was to find ways to change the cognition.

Kennedy has quite a mathematical description of what is going on: "It's all about networks. You've got a very complex set of relational networks that determines who you are, but also how you respond to the world. So changing that network becomes important."

In Beck's initial model for CBT, our behaviors are viewed in a very algorithmic way. A trigger acts as an input, which gets processed to produce thoughts, feelings, and behaviors, which then might trigger an action or output. Beck proposed CBT as a way to break down this algorithm into smaller pieces to identify the bug in the program, the faulty cognition. The behavioral part of the therapy consists of exercises that the therapist gives the client to prove to them that certain parts of the algorithm are erroneous. For example, a fear of spiders might be addressed by gradual short exposure to a spider to reveal that the patient's fears of the consequences are unfounded.

The striking thing is how quickly in certain cases awareness of the faulty cognition can lead to positive changes in behavior. Better thinking leads to better well-being. The fact that this can be achieved in eight one-hour sessions has led to the explosion of CBT and other such therapies as a shortcut to getting people back to work. The highly structured nature of the therapy means it can often be provided in different formats, including groups, self-help books, and even an app on your phone.

The shortcut is believed to be so effective that it became the backbone of the British initiative Improving Access to Psychological Therapies (IAPT), which began in 2008 and has transformed the treatment of adult anxiety disorders and depression in England. Economist Richard Layard persuaded the Labour government at the time that the amount of money that would be saved by getting people back to work would mean that the scheme would end up paying for itself. In 2009, £300 million was provided by the government over three years to train an army of more than 3,000 therapists. IAPT is today widely recognized as the most ambitious program of talking therapies in the world. In 2019 more than one million people accessed IAPT services for help overcoming their depression and anxiety.

Sometimes circumstances don't allow for anything but a very short intervention. But Kennedy pointed me toward data to support the effectiveness of a CBT model that relies on just three sessions. First proposed by Michael Barkham, it is called two-plus-one. Clients are seen for two one-hour sessions a week apart followed by a third session three months later. A growing body of research is showing that even this very short shortcut can be effective. For example, data published in *The Lancet* in 2020 revealed how such an intensive two-plus-one model resulted in meaningful reductions in psychological distress among female South Sudanese refugees in Uganda. As the researchers stressed in their paper, innovative solutions are required to provide mental health support at scale in such low-resource humanitarian contexts.

The other aspect of Kennedy's approach that resonated with me was the use of diagrams as tools for exploring new perspectives. One such diagram is the cognitive triangle. This is a diagram that helps the therapist and patient to understand the integrated nature of thoughts, feelings, and behaviors. Sometimes the triangle is drawn as a square, where feelings are divided into two corners: emotions and bodily sensations. The idea is that without intervention in the flow around this shape, thoughts will trigger feelings that lead to unhelpful behaviors that the patient would like to address, like a fear of going outside or a phobia of spiders. But by understanding and being aware of the cycle, it is possible to intervene earlier to change behavior. The diagram is like a map of the patient's mental terrain. By rising outside the network of thought, the patient is able to see that there are choices of different paths they might take.

Kennedy describes another diagram that she gives to therapists rather than patients to think about during sessions. "Imagine that you're the therapist and I'm the client, and we're both sitting on opposite ends of a seesaw balanced on a rope and the rope is stretched across the Grand Canyon. It's very important for us both to maintain this balance. I come along to therapy one day and I'm in a great mood because I've done my homework and I made these changes. So I've moved toward you on the seesaw, and then you naturally, being a very enthusiastic and caring therapist, will move toward me on the seesaw. But the next week I come along and I don't think I can do this anymore. I've had a terrible week. Nothing's worked. I just feel like giving up. I've moved away from you on the seesaw. Your instinct would be to move toward me. In which case we'll crash into the Grand Canyon. The harder you try, the more I'm going to resist you. And so what you need to do is to move away from me."

It's a fascinating image because Kennedy has changed the therapy into an equation that, like a seesaw, needs to be kept balanced.

For Kennedy and others, the evidence of the effective nature of these shortcuts is in the data, much of it collected by David Clark, professor of psychology at Oxford University. Tens of thousands of therapists send in data from their clients every week, and Clark has been collecting data for a decade. He places all this data in the public domain in order to promote transparency about mental health outcomes.

But sometimes cognition is not enough. Sometimes there is no shortcut that can replace the deep, protracted work one needs to undergo in therapy in order to rewire the brain. Kennedy acknowledges that formulaic therapies have their drawbacks. "CBT is all based on logic. But actually there is other stuff to therapy as well, about self-acceptance and attachment and being part of a family and part of the group and part of the world that comes with a good enough upbringing.

That, if you want to fix [it], you cannot do in eight sessions." As a result, CBT can sometimes just be seen as putting a Band-Aid on a gaping wound. It might stop the bleeding in the short term, but if you don't address the cause of the wound, then it will just reopen again at some point. How can you rewire the brain in eight one-hour sessions? Some therapists fear that CBT is sometimes cutting corners rather than being a genuine shortcut.

I think partners of therapists are always curious about what exactly is happening behind the closed door of a therapy session. It was one of the reasons I pulled *In Therapy* by psychoanalyst Susie Orbach off Shani's bookshelf. It turns out that this curiosity was one of the motivations for Orbach to write the book, which is dedicated to Orbach's own partner, Jeanette Winterson, "who has always wanted to know what goes on in the consulting room."

Orbach became famous for treating Princess Diana's eating disorder. As she explains in the book, therapy isn't just about training the mind and body to do something new, like play the cello or speak Russian. You've got to start by doing the much harder task of unlearning something.

"The essence of the human is the consequence of our long learning outside the womb. We don't arrive knowing how to walk and talk and think and feel. We apprehend how to do so in the context of the relationships which receive us. Those relationships, embedded in time and place and economic circumstances, then structure our mind, our feelings, our brains, our desires, our behaviors, and the way we are embodied."

Therapy can take so long because you have to address the basic ways your mind makes sense of the world. As Orbach puts it, "In therapy you don't just learn a new language to add to your repertoire, you relinquish unhelpful parts of the mother tongue and weave them together with the knowledge of a new grammar."

When I contacted Orbach to explore this idea more, she stressed this point. But she also acknowledged that there are still shortcuts that she uses in her sessions with patients. The interesting thing I learned when talking to Orbach is the role that patterns can play. The therapist spots patterns of behavior that correspond to previous case studies to help in plotting a course of action for the new patient in the room. But this needs to be balanced with recognizing each case as individual and unique.

"The kind of therapy I do is that you draw the lesson from an in-depth study of one person. That's Freud's legacy. It's the case study. It doesn't mean they fit, but it means maybe 50 percent of it fits. So that's a shortcut in a way if it's embedded in you, in your thinking, in your cognitive and your emotional repertoire as a therapist."

This is one of the fascinating tensions that exist in psychology. On the one hand, it borders on being a science, because there are things like case studies and because patients come in with particular behaviors. A medical doctor looks to match symptoms to previous case studies to be able to navigate the patient through the ailment based on the previous histories. Patterns of behavior can similarly provide the therapist a shortcut to understanding a patient in the same way that patterns help a mathematician to apply a previous methodology to solve a seemingly new problem. And yet the unique nature of each individual's psychology means that you never get replication. Each case is individual and requires personalized treatment. This is the art rather than the science of being a therapist.

"Therapy is a bespoke craft with each therapeutic pair or group creating novel circumstances to respond to," says Orbach. "One truth can open to another, which may shade what is first understood. The intricate constructions of the human mind shift during the course of therapy. Being a participant observer to the changing of internal structure and of the expansion of feelings is very satisfying. Seeing how defenses are used, and how they can be worked around and in time dissolve, has a beauty which is perhaps akin to the mathematician or physicist's experience of finding an equation elegant."

Orbach suggested to me that the way she approaches each new patient is not so dissimilar to the way that I, as a mathematician, will approach each individual new problem.

"If I do an assessment of a potential patient I've got a physical sense, maybe even a sort of geometric picture in my mind, of internal object relations, defense structures, emotional this, that, and the other. I've got so many things going on, but I don't even know that they are going on until I have to write it out. So that constitutes a shortcut, but then again that has come from the fact that I've been at it for forty bloody years."

As ever, here's that recurring theme—that shortcuts are hard-earned, requiring many years of work. I wondered what Orbach's thoughts were on the way CBT is being used in therapy as a shortcut. Orbach told me she is skeptical of this almost algorithmic approach to therapy.

"I don't believe in manualized therapy. Does that mean it's useless? No. Something is better than nothing. But are you supposed to be better within eight weeks or eight sessions? The trouble with a lot of [IAPT] is that they're often not therapists doing it and it's a highly skilled job." Indeed, some of the CBT treatment is even being delivered by AI therapists. Orbach doesn't believe you can reduce therapy to a formula to be followed. "Human subjectivity is not trivial. It is infinitely complex and beautiful."

CBT might have the power to construct frameworks that allow patients to be able to see certain patterns of thought and understand where they come from. With awareness, patients can then act to short-circuit these negative automatic thoughts. But for Orbach, this misses an essential quality of therapy, which is that these patterns often work at the level of ideas but not emotions. And this is key to why she believes they can't truly shortcut therapy. Emotions play a critical role in high-level cognition and consciousness. You can't change the latter without addressing things at an emotional level. Emotions create cognitive structures that develop over decades. For example, noted Orbach, "you have a defense structure, so you might understand, yes, I'm reproducing this particular behavior because it's embedded in me and that is how I understand, for example 'love means hate' or 'love means hitting' or whatever. I understood that, but the emotional component of that is unbelievably complex. And so of course, it's an aid, but basically it's . . ." At this point she let out a huge sigh. "It's not easy."

CHAPTER 8

THE PROBABILITY SHORTCUT

Puzzle: Which should you put your money on?
1. Throwing six dice and getting at least one 6.
2. Throwing twelve dice and getting at least two 6s.
3. Throwing eighteen dice and getting at least three 6s.

MODERN LIFE CONSISTS OF a whole sequence of decisions based on assessing a range of probable outcomes. Risk analysis is an integral part of how we negotiate our way through the day. There is a 28 percent chance of rain today; should I take an umbrella? The newspapers declare that I have a 20 percent increase in my chance of colon cancer if I eat bacon; should I stop with those bacon sandwiches? Is my car insurance too high given my risk of an accident? Is there any point in me buying a lottery ticket? What are the chances that my next throw of the dice when I'm playing a board game will send me down a ladder?

Many professions are faced with calculating odds to make crucial decisions. What is the chance of a stock rising or falling? Is the accused guilty of the crime, given the DNA evidence presented? Should patients be concerned by false positives in medical tests? Where should a soccer player aim for when taking a penalty? Negotiating an uncertain world is a challenging task. But finding a path through the fog isn't impossible. Mathematics has developed a powerful shortcut to help us navigate the uncertainties of everything from games to health, from gambling to financial investment: the mathematics of probability.

Throwing dice is one of the best ways to explore the power of this shortcut. This chapter's opening challenge was one that had exercised the famous seventeenth-century diarist Samuel Pepys. Pepys had a fascination with games of chance, but he was always rather cautious about risking his hard-earned cash on the roll of dice. In his diary entry on January 1, 1668, he writes how as he was returning home from the theater he stumbled upon "the dirty 'prentices and idle people playing." He recalls how a servant had taken him there as a child to see men throwing dice in an attempt to win their fortune, "to see how differently one man took his losing from another, one cursing and swearing, and another only muttering and grumbling to himself, a third without any apparent discontent at all." The friend with whom he had attended the theater, Mr. Brisband, offered to give him £10 to try his own luck, saying that "no man was ever known to lose the first time, the devil being too cunning to discourage the gamester." But Pepys refused, and fled back to his chambers.

At the time he'd witnessed the dice being rolled as a child, there were no shortcuts that would give him an edge. But in the intervening years all that had changed, because across the Channel in France, two mathematicians, Pierre de Fermat and Blaise Pascal, had come up with a new way of thinking that would give the gambler a potential shortcut to making money, or at least losing less money. It may be that Pepys had not yet heard of the breakthroughs Fermat and Pascal had made in wrestling the dice out of the hands of the devil and into the hands of the mathematicians. Today in the casinos across the world, from Las Vegas to Macau, the mathematics of probability that they initiated is the key to keeping the casinos in business at the expense of the "idle people playing."

What Are the Chances of That?

Fermat and Pascal had been inspired to come up with a shortcut after they'd heard about a challenge similar to the one that Pepys was contemplating. A mutual acquaintance, the Chevalier de Mere, wanted to know which was the better bet out of the following:

(A) Getting a 6 after throwing a die four times
(B) Getting two 6s if you throw two dice twenty-four times

The Chevalier de Mere was not in fact a knight at all, but a scholar by the name of Antoine Gombaud, who used the name to represent his views in the dialogues he liked to write. But the name stuck and his friends started calling him Chevalier. He'd tried to solve the conundrum of the dice by taking the long route and doing lots of experiments, throwing dice over and over. But the results were inconclusive.

Instead he decided to take the problem to the salon organized by a Jesuit monk named Marin Mersenne. Mersenne was something of a hub for intellectual activity in Paris at the time, receiving interesting problems and sending them out to other of his correspondents who he thought might have some clever insights on the problem submitted. He certainly chose well when it came to the Chevalier de Mere's challenge: Fermat and Pascal's response was to establish the shortcut of this chapter, the theory of probability.

It's not surprising that the long route was not really helping the Chevalier de Mere to decide which option was the best bet. Once Fermat and Pascal applied their new shortcut of probability to the dice, it turned out that option A occurs 52 percent of the time, while option B occurs 49 percent of the time. Play the game 100 times and the errors that creep into random games of dice will easily obscure this difference. Only after about 1000 games might the true pattern of play emerge. That's why this shortcut is so powerful—it saves you from having to do a lot of hard work repeating experiments that might still give you the wrong feel for the problem.

The interesting thing about the shortcut that Fermat and Pascal came up with is that it only really helps you gain an edge in the long term. It wasn't a shortcut that helped the gambler win in any individual game. That was still in the lap of the gods. But in the long run it made a huge difference. That's why this was great news for the casinos and less good news for the idle gambler hoping to make a quick buck on a single throw.

Back in London, Pepys wrote about the fascination at seeing the players on his walk home trying to throw a 7: "To hear their cursing and damning to no purpose, as one man being to throw a seven if he could, and, failing to do it after a great many throws, cried he would be damned if ever he flung seven more while he lived, his despair of throwing it being so great, while others did it as their luck served almost every throw."

Was the man particularly unlucky at not ever getting a 7? The strategy that Fermat and Pascal came up with for calculating the chances of getting a particular score with two dice was to analyze all the different ways that the dice could land and then look at the proportion of those possibilities that give the score 7. There are 6 ways the first die can land, which, when combined with the 6 different ways the second die can land, gives a total of 36 different combinations. Of these, 6 give you the score 7: 1 + 6, 2 + 5, 3 + 4, 4 + 3, 5 + 2, 6 + 1. Given that each of the different combinations is equally likely to occur, they argued that 6 out of 36 times the dice will land on a score of 7. It is actually the most likely score that you'll get with two dice. But there is still a 5-in-6 chance of not getting a 7.

With that in mind, how unlucky really was the gentleman Pepys saw who felt such despair at failing so many times to roll a 7? What are the chances that he failed to get 7 after rolling the dice 4 times? To go through all the different scenarios looks rather daunting, since there are $36^4 = 1,679,619$ different outcomes. But Fermat and Pascal come to the rescue, because there is a shortcut. To work out the chances of not getting a 7 in four rolls, it turns out that you simply have to multiply the chances on each roll: $5/6 \times 5/6 \times 5/6 \times 5/6 = 0.48$. This means the odds of not getting a 7 four times in a row are still only about even.

Conversely, this means that in four rolls of two dice there is a roughly even chance of seeing a 7. The same analysis proves that the odds of getting a 6 after rolling a die four times are also even. So the fact that the gentleman Pepys had observed hadn't got a 7 after four throws isn't that surprising. It's the same as not getting a head with one toss of a coin.

The fact that 7 is the most likely throw of the dice can be used to your advantage when playing many games involving dice, such as backgammon or Monopoly. For example, Jail is the most visited square on the Monopoly board. Combining this with an analysis of the likely throw of two dice means that after being on the Jail square, many players will visit the orange region of properties with greater frequency than anywhere else. So if you can grab the orange properties and stack them with hotels, you'll give yourself a crucial edge in playing the game.

A Cunning Shortcut: Consider the Opposite

Hidden in the calculation that Fermat and Pascal made is another clever shortcut often used by mathematicians. What if I'd started with the challenge of trying to calculate the probability of getting a 7 in four rolls of the dice? The thing is, clearly you don't arrive at that probability by multiplying the chance of getting a single 7 by four times. That calculation gives me the rare case of getting four 7s in a row. Instead I have to go through all the possible combinations of 7 that might occur. I need to calculate the chance of getting 7 on the first throw and no 7s after that, or no 7s on the first two throws and then two 7s on the last two throws—again, a mess of work. But here is the powerful shortcut. There is only one case I'm not interested in: failing to get any 7s. But this probability was easy to calculate. So instead of going at the problem head-on, look at the negative.

I find this a very powerful shortcut whatever problem I am working on. If tackling the problem head-on is too complicated, try looking at the opposite. For example, understanding consciousness is a tough scientific problem, but analyzing when something is not conscious can sometimes give you new insights into the more direct challenge. This is why analyzing patients in deep sleep or a coma can help scientists understand what it is that makes an awake brain conscious.

This shortcut via the negative challenge is key to working out the following problem: Each weekend in the United Kingdom, ten soccer matches take place in the Premier League. What proportion of them have two players on the field who have the same birthday?

At first glance that would seem to be a pretty rare occurrence. Maybe one out of ten? I think one's intuition on this is influenced by thinking that this question is the same as asking, "If I was playing soccer this weekend, what are the chances that someone on the field with me has the same birthday as mine?" There is about a 5 percent chance of that happening.

The point is that you are just thinking about pairing yourself up with each player on the field. But what about all the other possible pairings? It doesn't have to be you that shares a birthday. With that insight, the problem starts to get more complicated, and you begin to see that there are lots of different ways that people can pair up.

But by using the shortcut of looking at the negative problem, there is a much more efficient way to solve this problem. What are the chances that no one on the field has the same birthday? If I can calculate this, then subtracting this from 1 will give me the probability that there are two people who do share a birthday.

Get each person playing in our soccer match to come onto the field in turn. I run on first. Then the next player runs on. What is the chance that he has a different birthday than I do? The answer is 364/365. He just has to avoid my birthday: August 26.

Now the next player runs on. That person must have a birthday different from mine and also from that of the second player on the field. There are still 363 days to choose from, so the chances they don't have either of our birthdays is 363/365. With three of us on the field, the chances that none of us has the same birthday is 364/365 × 363/365.

Now I just keep going, bringing on all twenty-two players—and the referee, who is also on the field. Each time another person runs onto the field the number of possible birthdays that I have to avoid goes up. By the time the referee comes on, they have to avoid the twenty-two birthdays already on the field, so the chances of that are (365 – 22) / 365 = 343 / 365. The probability after all twenty-three people are on the field that none of them have the same birthday requires this calculation:

$$364/365 \times 363/365 \times 362/365 \times \ldots \times 344/365$$
$$\times \, 343/365 = 0.4927$$

I've calculated the opposite of what we want. Now I just need to flip this. The chances that there are two people with the same birthday is 1 – 0.4927 = 0.5073. Incredibly, it's more than likely that there is a birthday coincidence. Which means that on average, five out of the ten matches in the Premier League each weekend will see two people on the field with the same birthday.

Interestingly, the odds are probably higher than even, because there is evidence that professional soccer players are more likely to have a birthday in early autumn, in the months of September or October. Why? At school the impact of being born early in the school year means that you are likely to be physically more developed than someone like me who was born in August, and hence you will be stronger and faster and more likely to be picked for the school soccer team and to get experience playing the game. I remember vividly wondering why I never won races at school. And then one summer at a fair in our town I took part in a race that required you to run according to your age group. Because it was the summer, I hadn't yet had my birthday, while all the people in my school year group had already turned a year older; this meant I was up against the kids from the year below mine at school. I was totally shocked to find myself crossing the finish line first, for the first time ever, as I left the kids in the lower year behind.

But scrawny young me still had to make do with sitting in the library and becoming a math whiz!

The Shortcut to the Casino

Mathematicians are in high demand in Las Vegas because casinos are constantly looking for new shortcuts that can rig their games in the house's favor. Take craps, the game that evolved from the one Pepys observed. Betting on craps is quite a complicated business because of the dynamics of the game, but at any one time you can make a bet that the next roll of the dice hits a 7. I have already explained that on average this happens 1 in 6 times. And yet if you place $1 on this bet and win, the casino only pays you $4 on top of your dollar. To make this a fair game they'd need to pay out $5. This turns out to be one of the worst bets you can make at the craps table, as it gives the house a 16.67 percent edge. This is the profit (on average) that the casino makes each time a gambler makes this bet.

If you are absolutely insistent on betting on 7, then there is a better way to decrease the house's edge, which is to spread your bet three ways. Instead of placing one bet on getting a 7, you place three bets. One bet goes on the dice landing on 1 and 6. The second bet on 2 and 5, and the third bet on 3 and 4. This is called a hop bet. Although making these three bets is effectively the same as betting on a combined total of 7, the hop bet offers the chance of a better payout than you'd get if you bet straight on 7—the house only makes 11.11 percent profit (on average) every time you place a hop bet.

Every game in Vegas has been carefully analyzed to make sure that in the long run the house gets the edge, but if you want to gamble, you can use the tools that Pascal and Fermat developed to find the places where you have the best chance of losing money slower than anywhere else.

For example, in craps there is a bet where actually the house pays out according to the odds of winning. It is about the only place in the casino where the game hasn't been biased in the house's favor. Craps works by the player setting a target score and then throwing the dice. The target score needs to be one of the following: 4, 5, 6, 8, 9, 10. If the dice land on 2, 3, 7, 11, or 12 the game is over; 7 or 11 wins the game for the player, and 2, 3, or 12 is a losing score and is referred to as "crapping out." If a target has been set, then the aim is to get this score again with the dice before a 7 shows up.

The fair bet that you can make is to bet on this target appearing before the 7 appears. Suppose that a target of 4 has been set. If you bet $1 on seeing a 4 again before the 7, then the casino will pay you $2 on top of your $1 stake if this happens, giving you back $3. This is exactly the odds that this will happen. There are three ways to get a 4 and six ways to get a 7, so you will only win this bet 1 out of 3 times. This is a bet where the casino pays without taking a cut by stacking the odds in its favor. Knowing this doesn't give you a shortcut to making money, but at least it shows that you aren't giving money away. Betting on this scenario means that in the long run you should come out even.

Here's a little challenge for you. Let's move to the roulette wheel. You have $20. Your goal is to try to double your money. If you put money on red and it comes up, then you get double your money back. Which strategy is more likely to work? Strategy A: Put all your money in one go on red. Strategy B: Put down $1 on red on each spin of the wheel.

At first sight it might not seem to matter, but there is a slight twist to the roulette wheel. There are 36 numbers, half red, half black, but then there is also a 37th number: the zero, which is green. If the ball lands here you lose your money, whether you bet on red or black. This is the place where the house beats everyone. It looks pretty innocent, but the casino has calculated that this is their shortcut to profitability, at least in the long run. It means that there isn't an even chance of you winning your bet if you put your money on red. You have a slightly smaller than even chance: 18/37. Suppose that you make 37 bets of $1 on red and by a fluke the roulette wheel turns up each number once over the 37 spins. Then on 18 of them you win $1, but on 19 you lose $1, so at the end you've only got $36. It means that essentially every time you play, you pay the house $1/37 = $0.027 for each dollar you bet, giving the house an edge of 2.7 percent. The more you play, the more you pay.

In Strategy A the chance of you doubling your money in one go by putting down your $20 is 18/37 or 48 percent, just less than even. But if you play Strategy B, because you are paying for each $1 bet you make, it means that this strategy is gradually getting you further and further from your goal of doubling your money. In fact, there is only a 25 percent chance that this strategy nets you double your money in the long run.

Although Strategy A is the best bet, it does mean you have rather a short time in the casino. Strategy B might be a more fun evening, but you pay for your entertainment.

You might have heard that the place to go in the casino to get an edge over the house is the blackjack table. Mathematician Edward Thorp worked out in the 1960s that by studying the hands of the dealers and the other players, you can get an edge. The method is called card counting. In blackjack you are trying to get cards that add up to 21 or less and that beat the dealer's hand. If you go over 21, you go bust and lose. The key fact that makes counting cards work is that if the dealer's cards add up to 16 or lower, then the dealer will always take another card.

A deck of cards has 16 cards that have a value of 10 points (10, jack, queen, king). If you know that a lot of these cards are still in the deck, then it means that the dealer has a much higher chance of going bust if they have to take a card. Counting cards is a simple method of keeping track of how many high-value cards have already been played and how many are still left in the deck. If the deck is still loaded with high-value cards, then the casino is more likely to go bust, so it makes sense in this setting to bet high. Casinos in general try to minimize the effect of counting by using six to eight decks of cards for a game, rather than a single deck, but counting can still give you an edge. The film *21* dramatized the true story of a visit to Vegas by a team of mathematicians at MIT who implemented Thorp's shortcut. The nerdy mathematicians come off looking so sexy and cool that the film has probably done more for university admissions in mathematics than the combined recruiting efforts of all the math departments across the country.

At first sight this looks like a great shortcut to getting rich. The only trouble is that when I did an analysis of how long it would take you in reality to make much money from applying this strategy, it turned out that the time you would need to put in meant you would be earning less than the minimum wage. It seems that luck played a part in the MIT team's success.

Entry Fee

How much would you be prepared to pay to play the following game? I roll a die and pay you in dollars the number that appears on the dice. There is a one-in-six chance that you roll a 6 and win $6. Each of the other options also occurs with a 1-in-6 chance. Over six rolls I might make $1 + 2 + 3 + 4 + 5 + 6 = \$21$. So the average payout per roll is $21/6 = \$3.50$. If anyone offers you less than this to play, it would be worth playing, because you'll be the winner in the long term. Every time you play a game for money, it is sensible to assess what the average payout is likely to be, to determine whether the game is worth playing.

Although it was the correspondence between Fermat and Pascal that led to the discovery that you could apply mathematics to games of chance, the mathematics of chance really crystallized with the publication by Swiss mathematician Jakob Bernoulli of *Ars Conjectandi* (The art of conjecturing). Jakob was part of the Bernoulli dynasty that had backed Leibniz in the controversy over the calculus. It is here that you find the formula for the fair price you should pay for any game.

Suppose there are N possible outcomes. You win W_1 dollars if outcome 1 occurs. This happens with probability P_1. Similarly, outcome 2 occurs with probability P_2, in which case you win W_2 dollars. On average this game earns you $W_1 \times P_1 + \ldots + W_N \times P_N$ dollars each time you play. So if anyone offers you the opportunity to pay less than this to play the game, you're going to be the winner in the long run. For example, in the dice game there are 6 outcomes, the probabilities $P_1, \ldots, P_6$ are all 1/6, and the winnings $W_1, \ldots, W_6$ run from $1 to $6.

The formula seemed sound until Jakob Bernoulli's cousin Nicolaus, in an almost Oedipal act, came up with the following game. I toss a coin. If it lands heads, I pay you $2 and the game ends. If it lands tails, then I toss again. If the second toss is heads, I pay you $4. If it is tails, I toss again. Each time I toss, the payout doubles. So if I toss 6 tails followed by a head, I'll pay you $2 \times 2 \times 2 \times 2 \times 2 \times 2 \times 2 = 2^7 = \$128$. How much would you be prepared to pay to play Nicolaus's game? $4? $20? $100?

There's a 50 percent chance that you'll only win $2. After all, the probability that it lands heads first toss is 1/2. So $P_1 = 1/2$ and $W_1 = 2$. But you're hoping for a long run of tails followed by a head to get as big a prize as possible. The probability that you get a tails followed by a heads will be $1/2 \times 1/2 = 1/4$. But this time you win $4. So the second outcome has $P_2 = 1/4$ but $W_2 = 4$. As you keep going, the probabilities get smaller but the payout gets bigger. For example, 6 tails followed by a heads has a probability of $(1/2)^7 = 1/128$ but wins you $2^7 = \$128$.

If you stopped the game after 7 tosses, then you would lose only if there were seven tails in a row. Using Jakob's formula, the average payout would be $W_1 \times P_1 + \ldots + W_7 \times P_7 = (1/2 \times 2) + (1/4 \times 4) + \ldots + (1/128 \times 128) = 1 + 1 + \ldots + 1 = \$7$. It's worth playing the game, therefore, if anyone charges you less than $7 to play.

But here is the sting. Nicolaus is prepared to play the game indefinitely until a heads appears. You're a winner every time. So how much will you pay to play the game? There are infinitely many options now. The formula says that the average payout will be $1 + 1 + 1 + \ldots$, which is infinity dollars! If anyone offers to play this game with you, it's worth playing whatever the cost to play. If the entrance fee is over $2, then 50 percent of the time when a heads appears on the first throw you lose out. And yet in the long run, if you keep playing the game over and over, the math says you come out on top.

But why is it that most of us wouldn't pay more than about $10 maximum to play the game? This is called the St. Petersburg paradox after Nicolaus's cousin Daniel, who, while working at the Academy in St. Petersburg, came up with the first explanation of why no rational person would pay any price to play the game. The answer is what any billionaire will tell you: the first million you earn is worth so much more than the second million. The number you should put in the formula is not the exact amount you win but what that prize is worth to you. In this way the price to play this game will vary according to how you value the outcomes. Daniel's resolution goes far beyond just the curiosity of a mathematical game: it is essentially the foundation of modern economics.

To illustrate how again this shortcut to becoming a billionaire isn't quite the shortcut it appears to be, consider this question: If you could play a game a second, how long would it take to play 2^{60} games? This is the number of games you might expect to play to break even in the St. Petersburg game if the entry price was $60. The answer is over 36 billion years. The universe is at most 14 billion years old. That's another explanation for why most people wouldn't pay an arbitrary price to play the game.

Goats and Cars

A problem from an American game show from the 1960s called *Let's Make a Deal* had people all over the world, including professional mathematicians, up in arms about the best strategy. The game show had a final round that went a bit like this.

I've got three doors. Behind two of the doors there is a goat, and behind the other is a brand-new sports car. In the following analysis I am assuming the contestant wants to win the car rather than a goat.

The contestant gets to pick a door, say door A. Fairly straightforward so far—there's a 1-in-3 chance that the car is behind the door, right? But here comes the twist. The host, who *knows* where the goats are, opens one of the other doors and reveals a goat. Then he offers you the choice of sticking with your original door or swapping. What would you do?

Most people's intuition is that there are now two doors, so there is a 50-50 chance that the door they chose originally has the car behind it. According to this line of thought, if you switched at this point, it wouldn't change the odds of winning, and you'd kick yourself if in fact you'd actually chosen the right door at the outset. So most people stick.

But you actually double your odds of winning by switching. It sounds strange, but here's why. To calculate the probability of you winning, I have to go through all the different scenarios where you switch and then count how many give you a win.

Scenario A: The car is behind door A, the door you chose. You swap and then get a goat.

Scenario B: The car is behind door B. The game show host opens door C to reveal the goat. You swap to door B and win a car.

Scenario C: The car is behind door C. The game show host opens door B to reveal the goat. You swap to door C and win a car.

Each of these scenarios is equally likely. Yet in 2 out of 3 of them you win the car. If you stick, that strategy only gives you a 1-in-3 chance of winning. By swapping you actually double your chances of winning!

Don't worry if you don't quite get or believe that. This same explanation was published in a magazine, and more than ten thousand people, including hundreds of mathematicians, wrote in to complain that it was wrong. Even Paul Erdős, one of the greatest mathematicians of the twentieth century, got it wrong before he thought about it properly.

But if you're not convinced, how about this? Imagine that instead of three doors, there are a million doors. The game show host *knows* which one hides the prize. You've picked your door at random. There is a million-to-one chance you chose the right door. Now the game show host opens all but one of the other doors, revealing 999,998 goats. There are two doors left, the one you chose and the door that the game show host hasn't opened. Wouldn't you switch now?

The point is that you are being given information by the game show host when the other doors are being opened. He knows where the goats are. If I alter the setup, things can change. Suppose you are up against another contestant. You choose a door. They get to choose one of the remaining doors. Their door is opened to reveal a goat. What do you do now? Strangely, although it looks like you've got the same information—two doors, one with a car and one with a goat—this time it really is 50-50 that you get the car if you stick. The difference is that there is another scenario that has to be taken into account this time: if your door contains a goat, then the second contestant could have chosen the door with the car. That couldn't happen in the previous scenario because the game show host always reveals a goat. Imagine the million-doors scenario. Your opponent opens 999,998 doors, all with goats. They have been extraordinarily unlucky not to get the car, but you learn nothing about either of the remaining doors by this bad luck. It's now down to a 50-50 chance between the last two doors.

The Reverend Bayes

Probability seems to make a lot of sense when you are considering future events. If I am about to throw two dice, then 1 out of 6 possible scenarios sees the dice land with the sum of 7. The probability is the same for you or for me because we are putting a value on something in the future.

But what if you throw the dice, they land, and you cover the result from me? Then this throw has happened. It is in the past. It is definitely either 7 or not. There isn't anything in between. The trouble is that I don't know the answer. We still can assign a probability to this event (though some people find this controversial). The probability for you is different because you have information. My probability is quantifying my lack of knowledge of the situation. Suddenly probabilities depend on the amount of information each of us has. This is the quantifying of epistemic uncertainty.

As I get more information about the event, then my probabilities are going to change. But coming up with a mathematics to keep track of what value I should now be assigning the event given new data has resulted in different schools of thought.

For example, you throw a white billiard ball on a table randomly, secretly mark its position, then remove the ball. I have no information, so if I am asked to draw a line to guess where the ball might have landed, I might as well draw the line in the middle. But suppose I now throw five red balls, and you tell me between which red balls your original white ball landed. Suppose there are three on one side and two on the other. This is going to push my guess toward the side of the table with the two balls. But how far should I push the line with this new knowledge?

Some schools of thought say that you should draw the line 2/5 of the way along the table. But a controversial figure in the theory of probability, Thomas Bayes, suggested that actually it should be drawn at the 3/7 position, because there is some extra information that you are missing in your analysis—namely, the fact that before you knew anything, there was a 50-50 chance of a random ball being either on the left-hand side or on the right-hand side. Bayes throws these two extra balls into the mix in determining where to draw the line.

Bayes was a nonconformist minister preaching in Tunbridge Wells but was also something of an amateur mathematician. He died in 1761, but among the documents he left behind was a manuscript explaining these ideas of assigning probabilities to facts given only partial information. This manuscript was later published by the Royal Society under the title "An Essay Toward Solving a Problem in the Doctrine of

Chances." The ideas in this paper have become hugely influential in the modern practice of assigning probability values to facts we have limited information about.

In court cases, lawyers will try to put probabilities on the chances that someone was guilty of a crime. In fact, the person is either guilty or not, so this assignment of a probability is in some sense quite strange. It is merely meant to be a measure of our epistemic uncertainty. But according to Bayes, the probabilities change as we feed in new information that we've gathered. Often juries and judges don't understand the subtleties of Bayesian ideas, to the extent that judges have tried to dismiss these mathematical tools as being inadmissible in court.

The assigning of probabilities to events as a shortcut to understanding our uncertainty often gets misused. The trouble is that the general public doesn't have a good intuition about chance. That's why we have to resort to math as our shortcut so we don't get lost on the way. Take the following example.

We are told the person who committed a crime came from London. The person in the dock is from London. But that is pretty flimsy evidence. There is a 1-in-10-million chance at the moment that we have the criminal.

The jury is now told that DNA found at the scene matches the suspect's DNA, and that there is a one-in-a-million chance that the DNA actually belongs to someone besides the suspect. That sounds like certainty, and most people would convict on that evidence alone. But Bayes helps explain how we should update the probability of the suspect being guilty. If London has a population of 10 million people, it means there are 10 people in London whose DNA would match the DNA at the crime scene. So that's just a 1-in-10 chance that the person in the dock is guilty. What looked like a certain conviction is no longer so clear-cut. This case is quite easy to understand, but many uses of Bayes's theorem in court cases are much more complex, involving many different types of evidence that require computer software to be able to compute the probabilities of guilt. The trouble is that judges often don't understand the math and throw the expert evidence out of court, resulting in some horrendous miscarriages of justice.

Medicine is another area where probabilities are often cited, and again, if you don't understand how this shortcut is being used, then you can be led far from the destination you were hoping for. If you go for a scan for breast or prostate cancer and you are told that the scan is 90 percent accurate in detecting cancer, then most people are freaked out if they get a positive result. But should they be? It is important to have the extra data that only 1 in 100 patients is likely to have cancer. So out of 100 people tested, 1 will probably have cancer and the test generally gives a positive result. It's the false positives that cause the problem. Of the 99 that are tested, with a 90 percent success rate the test actually gets it wrong for 10 out of these 99 healthy patients. So if you test positive, then there is only a 1 in 11 chance that actually you are the one with cancer!

It is important to understand these numbers because the media love abusing them to make a scary story. To go back to a probability I mentioned earlier, let's say that eating bacon increases your chance of colon cancer by 20 percent. That sounds frightening. Should I give up those bacon sandwiches I love? If you look at the proportion of people who get bowel cancer, it's 5 in 100. If you eat bacon, that goes up to 6 out of 100. That's a less frightening way to couch this probability.

Pepys

What about Pepys's challenge of throwing dice to get a 6 that opened this chapter? What are the chances of getting at least one 6 in six throws? Again, the shortcut is to consider the opposite. The chance of not getting a 6 six times is $(5/6)^6 = 33.49$ percent. So the chance of getting at least one 6 is quite high, 66.51 percent.

What about throwing twelve dice and getting at least two 6s? Here as well there are too many different scenarios to consider, so let's use the trick of working out the opposite: the chance of getting (a) no 6s and

(b) exactly one 6. Case (a) is the same principle as before: $(5/6)^{12}$ = 11.216 percent. What about getting exactly one 6? There are twelve different scenarios here, according to which throw the 6 occurs on. The chance that the first throw is a 6 and the rest are not is $(1/6) \times (5/6)^{11}$. This is actually the same for all the other scenarios so the total probability is $12 \times (1/6) \times (5/6)^{11}$ = 26.918 percent. So the chance of getting two or more 6s in twelve throws is

$$100\% - 11.216\% - 26.918\% = 61.866\%$$

So option (a) is the better one to gamble on. If you do a similar case analysis on option (c), which is a bit messier than option (b), the odds get even worse at 59.73 percent.

Pepys had written to Isaac Newton about the problem in three letters he sent at the end of 1693. Pepys's intuition was that (c) was the more likely option, but Newton, applying the shortcuts of Fermat and Pascal, replied that the mathematics implied the opposite was true. Given that Pepys was about to stake £10 (the equivalent of £1000 in today's money), it was lucky that Newton's advice saved him from a shortcut to penury.

SHORTCUT TO THE SHORTCUT

At every step in our journey through life we encounter junctions with many different pathways leading into the distance. Each choice involves uncertainty about which path will get you to your destination. Trusting our intuition to make the choice often ends up with us making a suboptimal choice. Turning the uncertainty into numbers has proved a potent way of analyzing the paths and finding the shortcut to your destination. The mathematical theory of probability hasn't eliminated risk, but it allows us to manage that risk more effectively. The strategy is to analyze all the possible scenarios that the future holds and then to see what proportion of them lead to success or failure. This gives you a much better map of the future on which to base your decisions about which path to choose.

PIT STOP 8

FINANCE

E VERYONE IS LOOKING FOR A shortcut to great riches—purchasing a lottery ticket, placing a bet on a horse, founding the next Facebook, writing the next Harry Potter book, investing in the next Microsoft. While math can't guarantee your path to riches, it does still offer some of the best ways of maximizing your chances.

You might think that Isaac Newton, with all his mathematical techniques for optimizing solutions, would have been a successful investor, but after losing a great amount of money in a market collapse he declared, "I can calculate the movements of stars but not the madness of men."

But since Newton's day mathematicians have understood that there are clever shortcuts to making money in the markets. That is why the funds that are consistently doing well in both good times and bad are invariably staffed by mathematics PhDs. A great shortcut to finding the best fund for your savings would be to count up the number of mathematics PhDs on the fund staff. But how does knowing your mathematics help? Isn't it all driven by human whims and moods? Wouldn't a psychology PhD be more useful?

At the beginning of the twentieth century French mathematician Louis Bachelier proposed that investing in stocks is actually no different from betting on the toss of a coin. This was the first model that emerged of how prices vary over time. Bachelier believed that if we had complete knowledge of the markets, we would see that stock prices move up and down randomly. This behavior is known as the "drunkard's walk" because on a graph it looks like the path of a drunken person staggering down the street. Sure, overall prices might be affected by the outbreak of a pandemic. But once we have taken that knowledge into account, from that moment on a stock might then randomly head up or down.

Knowing this doesn't really give you an advantage. But it does if you recognize that the model is in fact wrong. Mathematicians in the 1960s realized that the randomness of a coin toss wasn't quite right because that would imply there was a chance that a stock might become negative in price. So a new model emerged that was still random but recognized that a stock has limits on its lowest value but could potentially be able to climb as high in value as it wanted.

The other way you can beat the market is if you can glean some hidden information in the prices. This can give you an edge. A bookie assigning odds for three horses running a race will ensure that you can't win money by betting on all three horses. But what if you know for some reason that one of those horses couldn't win? Then it is possible to spread the bet across the other two to guarantee a win.

Essentially this is the origin of the idea that Ed Thorp suggested in his 1967 book *Beat the Markets*. As I mentioned in Chapter 8, Thorp had already figured out how to get an edge at blackjack by counting cards. He'd even used a device to analyze the spin of a roulette wheel to make clever bets before he was thrown out of the casino for cheating. But his new idea would lead to the creation of the concept of the hedge fund. The key was finding a way to invest in two financial horses such that you profited regardless of which one was successful.

Thorp had discovered that certain financial products, called warrants, were overpriced—a bit like how a bet in the casino is overpriced to give the casino a house edge. Unfortunately, in the casino you can't bet that you'll lose, so it's impossible for a gambler to take advantage of this knowledge. But Thorp realized there was a way to exploit the overpricing of warrants using something called shorting the market. You borrow the expensive warrants owned by someone else with the promise of giving them back the warrants at a later date. Then you can sell the warrants and when the time comes to give them back, in general the price of the warrant will be lower than the price you sold the warrants at, making you a profit.

The only problem is that sometimes this won't be the case. The warrant could possibly go up over time, just in the same way that a bet at the casino can still be successful even if the house has an edge. And the trouble is that if the warrant does amazingly well, you could be liable for a big loss. But here was the clever hedge. A warrant is an option to buy a stock. If the warrant does well, then that's because the underlying stock has done well. So in tandem with selling the warrant you borrowed, you also buy some stock, so if you happen to be unlucky and the warrant does well (meaning that your bet on the warrant failed), you've got money from the stock doing well. It's not a guaranteed money earner, but Thorp understood that most of the time, whether prices go up or down, you profit.

The key was pricing this balancing act to your advantage, just like the gambler who spread bets on two horses knowing the third couldn't win. It's all about exploiting knowledge to your advantage. The casino does this, but the clever thing is that hedge funds had also spotted this shortcut to making money from the markets.

But mathematics isn't the only shortcut available to the investor. Helen Rodriguez, a friend of mine who is a very successful financial analyst, prepared for her career by studying history, not mathematics. And as it turns out, the skills of the historian offer a shortcut Helen routinely exploits to give her an edge in understanding when a company is under- or overvalued.

Helen specializes in high-yield bonds, also known as junk bonds, which are commonly used in buying and financing companies. If I buy a bond, I am lending a company money with the promise of a fixed rate of interest in return plus my money back at maturity. Junk bonds have a higher risk of default, and for that reason they also have higher returns.

"Here's your first shortcut: we use a rating scale of the credit quality of companies defined by the company's willingness and ability to pay," Helen explained. "It starts at AAA, for companies that pose little risk, right down to C for companies where there is little chance of recovering interest or principal. Bonds are high yield if the rating is below BBB–. If a company has a low rating, the interest has got to be higher to make it worth investing in. Hence the name high-yield bond."

You often hear in the news about a country's or a bank's credit rating being downgraded by a firm such as Moody's, which is one of the corporations that issue these ratings. These ratings are taking the multidimensional corporate world and attempting to project this complex messy collection onto a 1-dimensional line with C at one end and AAA at the top end.

Helen uses her historical tool kit to work backward. Can she look at the story of a particular company that's been given a particular credit rating and understand whether the bond is under- or overvalued? She is trying to squeeze out new information that might allow her to get an edge. It is quite an art to identify an aspect of a company's story that others have overlooked that gives new insights into the value of a bond. And that skill of seeing the bigger picture is one that historians often excel at.

"I was covering these 2500 German beauty shops, and the bonds were still stuck above par, and I was thinking, 'What a waste of time.' Then they had a bad quarter, which they blamed on terrorism in Germany. Then they had another bad quarter, and they still blamed terrorism, at which point I thought, 'This is a bit odd.' But the bonds were still stuck above par. So I started doing a bit of reading and discovered that Asian companies were coming into Europe and exploiting the internet to sell the same cosmetics, but maybe six months out of fashion at half the price. It's called the gray market. There were a couple of disruptor companies that were doing this and absolutely killing the German beauty market. So we sell at 103, and within a year the bonds are in the 40s. People hadn't woken up to this gray market."

Essentially Helen used a trick similar to Thorp's. She borrowed the bonds and sold them at $103 but was able at a later date to give back the bonds to the original owner from whom she'd borrowed them by buying them at $40, making a huge profit. She'd managed to use her hunch about the imminent collapse of the bond to her advantage. Often it's about seeing through a company's bravado about its worth.

"Companies will often not be as open as they should be about the fact that they have a problem. It's something as stupid as it's run by fifty-five-year-old men that don't understand teenage girls. It's often arrogance or vanity or just not understanding the way of the world. We've seen it massively in retail, the whole disintermediation and disruption that's happened because of the internet. It's shocking how late some of the managers of companies have seen it."

It seems to me that in some ways this would make it quite hard to come up with shortcuts, because you actually have to get to know a company quite intimately to get those kinds of insights. There's a lot of storytelling involved. Helen compared it to watching a soap opera.

"I've been following this one company, a Spanish gaming company. The restructuring took a year and a half, and literally every single day, I'd open the Argentinian newspapers and have to read them because Cristina Kirchner, the former Argentinian president, was using the gaming sector as a political football. That was the story that was driving the bond!"

It's the skills Helen learned training to be a historian that she believes give her the shortcut to tell the story of each company she is assessing. As she is watching the soap opera of each company play out, she needs to guess what's going to happen in the next episode before it airs. As Helen sees it, she needs to be able to synthesize a huge amount of information into something useful. This is what historians are good at. "It's like a puzzle you're trying to understand," she told me. "It's just like doing history. With ten different sources I need to come up with my narrative of what I think happened. That's why someone else might take the same sources and come up with a different narrative. You need this for there to be a market. You need the guy who thinks this is a great thing and you need another guy who thinks this is the end of the world. Then there's a trade."

Another shortcut that she taps into is at the top of my list of mathematical shortcuts: the power of spotting a pattern. "You can also find patterns in what happens to companies and what goes wrong, because they've all got the same problem, but maybe their segmentation of what they sell is a little bit different. I'm trying to spot the patterns in what's going to happen before everybody else and making that recommendation."

Having worked for many years for Deutsche Bank and Merrill Lynch, among others, making investments, Helen now works for Creditsights, providing independent corporate bond research for investors, like the analysis she did on the Spanish gaming company. So if you were reading this pit stop in the hope that I have some great shortcuts about how to invest your savings, my advice would be to combine the skills of the mathematicians together with the deep knowledge that someone like Helen has gleaned from her training as a historian. As Newton articulated, sometimes the best shortcut is to stand on the shoulders of giants.

CHAPTER 9

THE NETWORK SHORTCUT

Puzzle: Draw the following figure without taking your pen off the page and without running over a line twice:

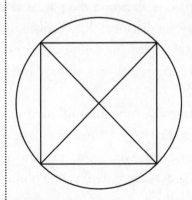

Figure 9.1.

O UR JOURNEY THROUGH THE modern world is increasingly mapped by networks. Systems of roads, railways, and flight paths allow us to get from one side of the planet to the other. A variety of apps offer the most efficient path through this intricate web. Companies such as Facebook and Twitter have stretched our social networks far beyond the inhabitants of our local village. The ultimate network that humanity spends hours navigating daily is the alternative world of the internet. Google rose to prominence with a shortcutting algorithm called PageRank, which helps users navigate around this web of nearly 2 billion sites. Although we regard the internet as a relatively new phenomenon, the first inklings of this network were actually hatched in the nineteenth century by my favorite shortcutter.

Gauss was a big fan of physics as well as mathematics and collaborated with Wilhelm Weber, one of the leading physicists in Göttingen, on numerous projects. Gauss even came up with a way to shortcut having to walk from the observatory in Göttingen down to Weber's lab. Instead of meeting in person, he rigged up a telegraph line between the two. It stretched across the roofs of the town for a distance of 3 km. Gauss and Weber had understood the potential of electromagnetism to communicate over a distance. They cooked up a code where each letter was represented by a sequence of positive and negative electrical pulses. This was in 1833, several years before Morse came up with a similar idea.

Gauss thought the idea was something of a curiosity, but Weber saw the import of such a technology: "When the globe is covered with a net of railroads and telegraph wires, this net will render services comparable to those of the nervous system in the human body, partly as a means of transport, partly as a means for the propagation of ideas and sensations with the speed of lightning." The rapid spread of the telegraph makes Gauss and Weber the grandfathers of the internet. Their collaboration is immortalized in a statue of the pair in the city of Göttingen.

Today this network, as Weber predicted, extends well beyond the few kilometers of wire that the two scientists trailed across the roofs in Göttingen. Indeed, it is so complex that finding shortcuts across networks has become one of the central subjects of modern mathematics. These networks can be made not just of wires but also bridges, as I explored on a recent trip to Russia.

"Read Euler. Read Euler. He Is the Master of Us All"

When I flew to Kaliningrad a few years ago I made sure I got a window seat for the short flight from St. Petersburg. I was on a pilgrimage to a city that is home to one of the stories that every mathematician is brought up on, one of the cleverest shortcuts dreamed up in the history of mathematics.

As we came in for a landing in Kaliningrad, a small exclave of the Russian federation cut off from mainland Russia by Lithuania and Poland, I could see the river Pregel running through the city. The river has two branches that join at Kaliningrad from where it flows west to emerge at the Baltic. There is an island in the center of the town around which the two branches run. It is the bridges connecting the banks of the rivers and this island that are at the heart of the mathematical story for which the city is famous.

The story dates back to the eighteenth century, when the town had a different name: Königsberg, birthplace of Immanuel Kant and the famous mathematician David Hilbert. During the eighteenth century, when Königsberg was part of Prussia, there were seven bridges spanning the Pregel. It had become a Sunday afternoon pastime among residents of the city to see whether they could find a way to cross all the bridges once and once only. But however hard they tried, they always found that there was one bridge that they couldn't get to. Was it really impossible, or was there some way the residents hadn't tried that would have allowed them to cross all seven bridges?

For the residents of Konigsberg, there didn't seem to be any way to avoid the hard work of trying every possible route around the bridges until they had exhausted all the possibilities. There was always a sneaky feeling that they might have missed a clever way around the bridges that meant the challenge was possible.

It took the arrival of one of my mathematical heroes, Leonhard Euler, to resolve the puzzle once and for all: it was impossible to

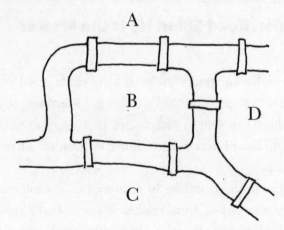

Figure 9.2. The seven bridges across the river Pregel in
eighteenth-century Königsberg

cross all the bridges once and once only. To figure this out, Euler discovered a shortcut that avoided having to try every route around the bridges.

I introduced Euler in Chapter 2 when I revealed the extraordinary formula he came up with linking five of the most important numbers in mathematics. "Read Euler. Read Euler. He is the master of us all," wrote Pierre Simon Laplace, one of France's foremost mathematicians, about Euler's importance to mathematics. Most mathematicians would agree, ranking him alongside Gauss as one of the greats. Even Gauss was a fan: "The study of Euler's works will remain the best school for the different fields of mathematics and nothing else can replace it."

Euler's contributions range far and wide, including the shortcut to solving the challenge of the bridges of Königsberg, which he first learned about when he was a professor at the academy in St. Petersburg. Euler was not a native of St. Petersburg but had traveled to the city from his hometown, Basel, where he'd been unable to find a job as a mathematician. Evidently all the positions in mathematics had been filled. Strange that such a small city should be so oversubscribed with mathematicians. Even stranger was that all the mathematicians came from the same family: the Bernoulli family.

Basel couldn't even accommodate all the Bernoullis. Daniel Bernoulli had already decamped to St. Petersburg, and it was his invitation that secured a position for Euler at the academy. Before Euler set off, Daniel sent him a letter listing all the Swiss creature comforts they were missing in St. Petersburg: "Please bring fifteen pounds of coffee, one pound of the best green tea, six bottles of brandy, twelve dozen fine tobacco pipes and a few dozen packs of playing cards."

Weighed down with all the provisions that Daniel had requested, Euler needed seven weeks to reach St. Petersburg from Basel. Traveling by boat, on foot, and by post wagon, he eventually arrived in May 1727 to take up his position.

The Bridges of Königsberg

At first, the problem of the bridges of Königsberg was for Euler little more than a bit of light relief from all the complicated calculations he had been engaged in. In 1736 he wrote a letter to Giovanni Marinoni, court astronomer in Vienna, describing what he thought of the problem:

> This question is so banal, but seemed to me worthy of attention in that neither geometry, nor algebra, nor even the art of counting was sufficient to solve it. In view of this, it occurred to me to wonder whether it belonged to the geometry of position, which Leibniz had once so much longed for. And so, after some deliberation, I obtained a simple, yet completely established, rule with whose help one can immediately decide for all examples of this kind whether such a round trip is possible.

The important conceptual leap that Euler made was that the actual physical dimensions of the town were irrelevant. What was important was how the bridges are connected together. The same principle applies to the map of the London Underground: it is not a physically accurate map, but just retains information about how the stations are connected. If one analyzes the map of Königsberg, then, just as locations in London become points on the Underground map, the four regions of land connected by the bridges could each be condensed to a point with the bridges represented by lines connecting the points. The problem of whether there was a journey around the bridges was then equivalent to being able to draw the resulting picture without taking your pen off the paper and not running over any line twice.

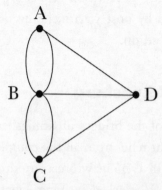

Figure 9.3. The network diagram of the bridges of Königsberg

So why was this impossible? Although Euler probably never drew this graphical picture of Königsberg explicitly, his analysis shows that if a journey was possible, each point visited in the middle of a journey must have one line into it and one line out. If you visited that point again, it would be on a new bridge into it and a new bridge out of it. And so there must be an even number of lines attached to each point. The only exceptions to this were at the beginning and end of a journey. Where you set off from would have one line coming from it, as would the end of the journey, which would also have one line running into it. For a journey on any graph to be possible, no more than two points— the beginning and end points—can have odd numbers of lines emerging from it.

But if you look at the plan of the seven bridges of Königsberg, each point has an odd number of bridges connected to it. With so many points with an odd number of bridges sprouting from them, it meant that a path around the city crossing each of the bridges only once was impossible.

This is one of my favorite examples of a shortcut. Instead of trying lots of different ways to trace out a route around the map, this simple analysis of the number of points with an odd number of bridges emerging reveals immediately that the route around the map isn't possible.

The beauty of Euler's analysis is that it doesn't just apply to Königsberg. In any network you draw of points joined by lines, if the number of edges emerging from a point is always even, then Euler proved that there will always be a path running over all the lines once and once only. Also, if there are precisely two places with an odd number of edges, then a path is possible where these points will be the beginning and end of your journey. It doesn't matter how complex the map—this simple analysis of the tally of odd-number nodes provides a shortcut to knowing whether the network is navigable or not.

Königsberg only had seven bridges, but recently mathematicians in Bristol applied Euler's shortcut to the forty-five bridges that span the complex system of waterways that run through that city. Whereas Königsberg has one island, Bristol has three: Spike Island, St. Phillips, and Redcliffe.

It is far from clear at first that a walk around all forty-five bridges would be possible, but using Euler's shortcut you can see that the number of odd-number nodes in the map charting the way the land was connected by bridges was small enough that a path had to be possible. A route for Bristol's walkable bridges was first devised in 2013 by Dr. Thilo Gross, a former reader in engineering mathematics at Bristol University: "Having found the solution I naturally had to walk it. The first bridge-walk took 11 hours and was about 53 km long."

This shortcut actually helped me during a set of psychometric tests that I was required to take for a job application when I was younger.

The test included a series of networks, and the task was to draw the networks shown without taking my pen off the paper and without running over a line twice. The implication of the task was that this was possible, and it looked like they were testing my ability to complete each task. What the task was really testing was the honesty of the applicant, because one of the three networks was impossible to draw. Like the map of the bridges of Königsberg, there were more than two nodes with an odd number of lines emerging from the point.

My essay next to the challenge explaining why this task was impossible, thanks to Euler's shortcut, didn't go down too well. I didn't get the job.

Human Heuristics

Euler's great insight was to home in on the essential quality of the map of the city that was important for solving the problem. It wasn't important how far you had to travel or what the bridges looked like. The art here was to throw away all the extraneous information and retain the quality of the map that was important for navigating the journey. This idea of throwing away unimportant information is key to many shortcuts. It's the idea behind humans' use of heuristics: the process by which we make problems less complex by ignoring some of the information that's coming into the brain, either consciously or unconsciously. We often have to make decisions based on limited time or mental resources, so we have to find efficient ways to pick out aspects of the problem that contribute to a solution and don't use up precious mental space unnecessarily.

In their groundbreaking work, psychologists Amos Tversky and Daniel Kahneman identified three key strategies that we use as mental shortcuts to making decisions. We use the idea of patterns between different events, which they called *representativeness*. This is certainly a quality that I exploit in mathematics to shortcut rethinking a problem. The second strategy is called *anchoring and adjustment*. This is a process that starts with an initial piece of information that we understand or know, the anchor, which other situations are then judged against. The final strategy is the *availability* heuristic, which uses our local knowledge to judge a more general situation.

Clearly these last two are very prone to produce bias because in general we don't have very good anchors or terribly representative local knowledge. In his hugely influential book on the limitations of human heuristics, *Thinking, Fast and Slow*, Kahneman gives examples of how just mentioning a number before you ask a question can skew people's estimates terribly. For example, mentioning the years 1215 and 1992 contaminated people's estimates for the first year that Einstein visited the United States (1921), dragging the estimates lower or higher than those participants who were asked the question without an anchor, even though the anchoring dates were clearly unrelated to the question being asked.

The mathematical shortcuts that we have come up with over the centuries are an attempt to override the evolutionary shortcuts that can fail as our questions become more complex. These heuristics might help us to navigate our patch of the savannah, where things are less likely to vary too much, but they are not helpful in trying to understand universal truths.

The key to a good heuristic is to understand, as Euler did in Königsberg, that the nature of the bridges, the distances involved, and the geography of the city are not relevant to the problem. Only the way the landmasses are connected is relevant to solving the challenge.

Once I arrived in Kaliningrad I was intrigued to see on my visit to the modern city how many of the seven bridges were still standing. The city of Kaliningrad suffered devastating bombing by the Allies during World War II. Being an important port on the Baltic, it was a strategic location for the German fleet. Much of the historic city was leveled, including the famous university situated on the island at the heart of the town where Kant and Hilbert had learned their academic trade. So how did the bridges fare?

Three of the prewar bridges were still there. Two of the bridges had vanished completely. The two remaining bridges had been bombed during the war but had subsequently been rebuilt to carry a huge highway through the city.

In addition, some new bridges had appeared alongside the historic bridges. A railway bridge that I discovered pedestrians can cross over joined the two banks of the Pregel out to the west of the city, together with a new footbridge called the Kaiserbridge. There are once again seven bridges, but now in a slightly different arrangement from the eighteenth-century bridges that Euler analyzed. The beauty of Euler's shortcut is that it applies whatever the number and arrangement of the bridges. So my immediate thought was to see whether a journey around today's bridges might be possible.

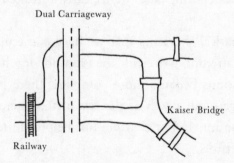

Figure 9.4. The seven bridges of twenty-first-century Kaliningrad

Euler's mathematical analysis showed that if there were exactly two places with an odd number of bridges emerging from them, then a path would always be possible: you start at one of the odd-numbered points and end at the other. Checking the plan of today's bridges of Kaliningrad reveals that such a journey is possible. Starting on the island in the middle of the city, I set off excitedly on my pilgrimage around the seven modern bridges of Kaliningrad.

The story of the bridges of Königsberg is like a mathematical fairy tale that all mathematicians will have been told at some point. But it is also the beginning of a very important branch of mathematics that is highly relevant to our digitally connected world: network theory. And the development of shortcuts around complex networks such as the internet has made some mathematicians a lot of money.

Internet Shortcuts

There are over 1.7 billion websites on the internet today. Yet despite this extraordinary number, the Google search engine still manages to quickly find the information you want to retrieve. You might think that this is the result of huge computing power, and that is certainly part of the equation. But it's the way Google searches that has made it such an indispensable tool.

In the past, search engines would look for those websites that mentioned your search term the most times. If you were looking for biographical details about Gauss's life, then searching on the term *Gauss biography* would bring up the sites that mentioned these two words most often.

But if I wanted to spread some false biographical details about Gauss, by loading the metadata of my site with many copies of the words *Gauss* and *biography* I could ensure that my false news site hit the top of the list. Just searching the internet using a word search did not provide a powerful way to find the sites you were after.

Two Stanford graduate students, Larry Page and Sergey Brin, working out of a garage in Menlo Park, came up with a much more robust means to question the internet to find the best way to rank which biography of Gauss to put at the top of a search request. They decided to exploit a clever tactic: to use the internet itself to tell it which pages were the most important. The idea was that the relevance of a website could be judged by the number of other websites that link to it. A legitimate page detailing the biography of Gauss would likely be linked to by other websites interested in the topic.

But if the importance of a website is simply judged by the number of links from other websites, then there would be an easy way for me to hack my false website to the top of the list. By making thousands of spoof websites and linking them to my Gauss biography page, I could seemingly make my page look the most important.

Page and Brin had a strategy to stall such a hack. A website would climb high in the rankings only if the websites that linked to it were also highly regarded. But hold on. This sounds rather circular. I need to know which of these sites linking to my Gauss biography sites are highly valued. But they get their value from high-value sites linking to them. I seem to have gotten into an infinite regress.

The way to resolve this was to regard all websites as having equal status at the outset. I start by giving each website a score of 10 stars. But now I proceed to redistribute the stars. If a website links to five other websites, then I give each website two of its stars. If it links to just two websites, then each website receives five stars. Although the website has given away all its stars, it hopefully is linked to by other websites that will give it some of their stars.

By continuing to redistribute stars from one website to another, after many iterations I start to see the dominant websites collecting more and more stars. Simply being linked to by my thousand spoof websites will be revealed for the hoax it is. After one round, my thousand sites are starless and can no longer help maintain the value of my false site. Very quickly my site gets drained of its stars and plummets down the list of sites that the algorithm is valuing. There is a little more work one has to do to implement this idea, but this is the essence of the way Google ranks websites.

The trouble is that it takes time and computing power to analyze how the stars flow around the network. But then Brin and Page realized that there was a shortcut to work out the ranking. As undergraduates they'd been taught what at first sight seems like a rather esoteric bit of arcane mathematics, called the eigenvalue of a matrix.

What this mathematical tool does is to identify in different dynamic settings certain parts of the system that remain stable. It was first used by Euler in the context of a rotating ball. If you take a globe with the countries of the earth painted on the surface, then no matter how you might twist and turn the globe in your hand, it is possible to take the final position of the globe and to fix two antipodal points such that a rotation around the axis through these points returns the globe to its starting position. Essentially it means that every possible rearrangement of the globe can be realized by a simple rotation through some axis.

The eigenvalue of a matrix provides both a proof that such an axis of rotation always exists and a method for finding the two stable points through which the axis runs. What is remarkable is how this technique allows us to identify points of stability in so many different dynamic settings. For example, eigenvalues of a matrix are central to identifying the stable energy levels in a quantum system. They are also key to picking out the resonant frequencies of musical instruments.

Brin and Page realized that they were also the secret to identifying how the stars would stabilize once they have been distributed around the network. So instead of running an iterative process waiting for things to reach an equilibrium, the eigenvalue of a matrix was a smart shortcut to calculate the page rank of any website on the internet.

Although my attempts to elevate my fake Gauss biography site were thwarted, it is still important for companies to understand how Brin and Page's shortcut works. There are things a company can do to make sure that the Google shortcut charts a path through the company's website. Small perturbations to the Google algorithm can see the shortcut alter course slightly, resulting in your website dropping down the rankings. You need to know what changes to make to get your website back on the pathway.

Social Shortcuts

Sometimes the challenge is how to get from one point in the network to another point via the shortest path possible. Are there cunning shortcuts across the network? Take the network of social connections across the population of the planet. If I choose two random people, how short a chain of friendship connections can I find to get from one to the other? The surprise is that it takes very few.

This question was first posed in a short story entitled "Chain-Links," written by Hungarian author Frigyes Karinthy in 1929. In it, the central character speculates that this network has amazing shortcuts across its chain of links:

> A fascinating game grew out of this discussion. One of us suggested performing the following experiment to prove that the population of the Earth is closer together now than they have ever been before. We should select any person from the 1.5 billion inhabitants of the Earth—anyone, anywhere at all. He bet us that, using no more than *five* individuals, one of whom is a personal acquaintance, he could contact the selected individual using nothing except the network of personal acquaintances.

It took just over thirty years for this fictional game to be put to the test. In a famous experiment conducted by American psychologist Stanley Milgram in the 1960s, a stockbroker friend of Milgram's living in Boston was chosen as the target. Milgram decided to choose two cities in America that he felt were furthest both geographically and socially from the target in Boston: Omaha, Nebraska, and Wichita, Kansas. Letters were then sent to random people living in these towns with instructions to forward the letter on to the named stockbroker. The catch was that no address was provided. If they did not know the named target, then they were asked to forward it to a friend in their network of acquaintances who they believed might be better placed to forward the letter.

Of 296 letters that got sent out, 232 never reached their target. But of those that did, on average it took six forwardings of the letter to get from the original recipient to the target. There were indeed five individuals between the start and end of the chain.

This experiment led to the famous phenomenon called six degrees of separation. The phrase was popularized by John Guare in his play of the same name. As one of the characters declares toward the end of the play:

> I read somewhere that everybody on this planet is separated by only six
> other people. Six degrees of separation. Between us and everybody else
> on this planet. The president of the United States. A gondolier in Ven-
> ice. Fill in the names. It's not just big names. It's anyone. A native in a
> rain forest. A Tierra del Fuegan. An Eskimo. I am bound to everyone
> on this planet by a trail of six people.

In our digital age we are more interconnected than ever, and it is a network we can explore much more easily than forwarding letters via the US postal system. In 2007 a data set of messages composed of 30 billion conversations among 240 million people revealed that the average path length between users was indeed 6. A paper published in 2011 found that across Twitter you can connect any two tweeters with a chain that is on average just 3.43 users.

Why do social networks have these shortcuts? It's certainly not true of all networks. Just arrange 100 nodes on a circle and join only those that are next to each other. It requires 50 handshakes to get from one side of this network to the other. A network in which you can move between any two points with only a small number of connections is called a *small world*.

It turns out that an extraordinarily large number of networks, not just our social and internet connections, are examples of small worlds. The neural connections in everything from the body of a *C. elegans* nematode, with 302 nodes, to the human brain, with its 86 billion neurons, seem to be examples of small-world networks. This allows one neuron in the system to quickly communicate with any other neuron via a short number of synapses. Electrical power grids are small worlds, as are airport networks and food webs. What is it about these networks that makes them small worlds?

Two mathematicians, Duncan Watts and Steve Strogatz, discovered the secret, which they published in a paper in *Nature* in 1998. If you take a set of nodes and create local links between those that are close to each other, then typically you get a picture like our circle, with long journeys that are necessary to connect randomly chosen nodes across the network. But what Watts and Strogatz discovered is that it required just a few global links across the network for shortcuts to appear. It's like everyone in Boston knows each other, but then there's someone in Boston who happens to have an aunt living in Kansas who will provide the way of connecting these local neighborhoods more globally. In the *C. elegans* nematode you can see the same architecture: the neurons are arranged in a circle but across the circle you see links connecting distant neurons. It seems that the human brain has a similar architecture—lots of local connections with a few long synapses linking disparate bits of the brain.

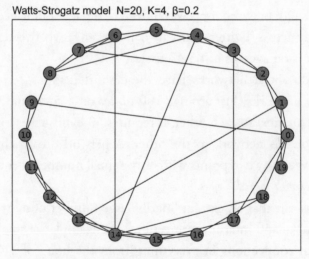

Figure 9.5. Example of small-world network

Airport networks work in a similar fashion, with a few airports that act as hubs connecting the world with long-range flights. And then within a region there are lots of short-haul flights that take you from the hub to the local destinations.

Using their mathematical model, Watts and Strogatz were able to show that in a network with N nodes where each node has K

acquaintances linked in this local-global manner, the average path between two randomly chosen points in the network is given by the
formula

$$\log N/\log K$$

where log is the logarithm function cooked up by Napier to shortcut
calculations. Put N equal to 6 billion, connect them up with 30 acquaintances each, and the number of degrees of separation comes out
at . . . 6.6.

If you are building a network, be it social, physical, or virtual, then
often you will want there to be shortcuts across the web of connections.
But now we know how to cook up such systems. If you want to create
a network with the small world's characteristic of amazing shortcuts
from one end to the other, it seems that adding this randomly selected
batch of global connections is what does the job.

Gauss's Brain

When Gauss, the great shortcutter, died in 1855, he left his brain to
science for investigation. His friend and colleague Rudolf Wagner, a
physiologist at the University of Göttingen, took on the task of dissect-
ing the brain to explore if there was anything special about it that might
have made Gauss so adept at finding mathematical shortcuts. It was
part of a larger project undertaken at the university to understand
whether there was any special architectural difference between elite
brains and those of the general public. Instead of crude measures such
as volume or weight, he proposed that Gauss's brain was more highly
convoluted in the cortex than a typical brain.

His work was complemented by a set of copper engravings and
lithographs that one of Wagner's team made. More recently with the
aid of modern high-resolution fMRI a team in Göttingen has indeed
confirmed a rather rare connectivity between two regions of the left
hemisphere of Gauss's brain. However, the team had to contend with a

strange mixup that happened in the collection. It turns out that for years what had been considered Gauss's brain was in fact that of another of Göttingen's elite, Conrad Heinrich Fuchs, who died the same year as Gauss. It seems that the specimens were mixed up after Wagner had made his analysis and the diagrams were drawn. It was only when the team came to compare the fMRI scans with the original drawings that the mixup was identified.

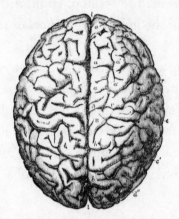

Figure 9.6. Gauss's brain

Göttingen's nineteenth-century project to understand the different architecture of the brains of elite thinkers continues to this day. At the University of Louisville, the anatomy department has been studying the brains of deceased scientists (or "supernormals," as they are called by the lab). Professor Manuel Cassanova, who has been leading the investigation, has detected structural differences in the brains of scientific specialists.

It seems that an abundance of short, local connections gives rise to brains that specialize in focused modes of thinking. These are individuals who more easily tap into the power of single regions in the brain. In contrast, brains with long connections that link diverse regions of the brain help create new ideas and out-of-the-box thinking.

It's interesting that this seems to correspond with a dichotomy that has emerged in different styles of thinking. "The fox knows many things but the hedgehog knows one big thing," asserts a fragment attributed to the ancient Greek poet Archilochus. It was the springboard for the foxy philosopher Isaiah Berlin's essay that seeks to divide thinkers into two categories. Foxes tap into a broad range of interests—a horizontal thought process. The hedgehogs think deeply—a thought process that runs vertically, perpendicular to the foxes' thought process. Foxes are interested in everything. The hedgehogs are single-minded in their obsessions.

If the abundance of short-range connections characterizes the hedgehog and a greater number of long-range connections characterizes the fox, wouldn't a brain that could combine lots of these short connections with an abundance of long links give rise to someone who can combine the skills of the fox and the hedgehog? That would be ideal, but the fact is that the wiring within the brain requires both space and metabolic activity. And as long as you have the constraint of the geometry of the skull, it is impossible to maximize both.

But there is an alternative: collaboration. Gauss collaborated with Weber to produce the first telegraph wire, which eventually would give rise to the modern internet. It is through sharing our specializations—creating these long-range connections between brains that can specialize—that we have the potential to get something new and exciting emerging. The low-hanging fruit consists of learning the language spoken by someone outside your discipline and applying it to the problems in your own field. This is why whatever your field of work, learning the ideas from another field might enable you and someone in the other field to collectively find a shortcut to the other side.

Perhaps the perfect fusion of fox and hedgehog is the collaboration of human and machine. Although my book is meant as a celebration of the distinctly human trait of sniffing out the shortcut, perhaps I shouldn't be so dismissive of what the machine can offer. While the machine can use brute force to calculate faster and further, ultimately it is in combination with the human ability to find clever shortcuts that both together can reach destinations that are out of reach of each individually.

Solution to the Puzzle

This is the challenge I was offered in my psychometric job application. Thanks to Euler's shortcut, I know that it is impossible to draw this figure because there are more than two nodes that have an odd number of lines emerging from that point. However, there is a way to draw this if you use a trick. Take a piece of paper and fold the bottom quarter over. Start by drawing a square from the top left corner, but make sure that the lower line runs across the paper you folded. Once you have completed the square, fold the page back to leave three lines of the square with your pen on the top left corner. If you now analyze the diagram that is left to draw, it passes Euler's test.

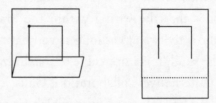

Figure 9.7. The trick to drawing the figure: fold the paper

SHORTCUT TO THE SHORTCUT

....

Networks are everywhere: the structure of a company, the electrical wiring of a computer, the interdependence of different stock options, transport networks, the cellular interactions in our bodies, the relationship between characters in a novel, our social networks. Whenever you've got a set of objects and connections between those objects, you've got a network. It is worth analyzing whatever structure you are trying to understand to see if there is a network hidden inside. Because once you have identified a network, mathematics is ready with shortcuts to help you navigate its architecture—tools to identify the most important nodes in the network, strategies to transform networks into small worlds with fast paths from one end to the other, topological maps that throw away extraneous information and help you see what's really going on.

NEUROSCIENCE

O FTEN THE BEST IDEAS SEEM to come from nowhere. It's as if *not* thinking helps the brain to find shortcuts to answers. The philosopher Michael Polanyi believed that this tacit thought process, where the brain taps into subconscious, unarticulated arguments, was key to the power of human thought. He summed up his thesis with the phrase: "We can know more than we can tell."

This is certainly my experience of creating mathematics—that sense of "seeing" the answer even though I'm not quite sure why I feel it's right. It's how I manage to come up with conjectures about what I think the lay of the mathematical land is like. I sense the existence of a far distant peak without quite knowing how to navigate a path to get there.

Many mathematicians have talked about the flashes of insight, the way the brain seems to throw an idea into our consciousness. The brain works away subconsciously, and once it hits a solution it then knows to bring that up into the conscious arena. I have had these flashes of insight, which are then followed by the often painful task of piecing together the logic of how my subconscious came to this conclusion.

Henri Poincaré describes a famous occasion when he'd been working away on a problem, unable to make any headway. It was only when he was away from his desk, letting his mind idle, that he had a sudden realization of how to solve the problem, which came as he stepped onto a bus in Paris: "At the moment when I put my foot on the step the idea came to me, without anything in my former thoughts seeming to have paved the way for it, that the transformations that I had used to define the Fuchsian functions were identical with those of non-Euclidean geometry."

Alan Turing had a similar experience when he was working on his idea of Turing machines. Turing used to enjoy running along the banks of the river Cam in Cambridge as a way of relaxing after working hard in his rooms. It was while lying on his back in the meadow near Grantchester that he understood how the mathematics of irrational numbers could be used to show why his Turing machines had limitations in what they could compute.

To learn more about solving a problem by not thinking about it, I decided to contact a neuroscientist, Ognjen Amidzic, who has been exploring brain function in people engaged in their particular fields of expertise.

Amidzic hadn't intended to become a neuroscientist. His dream had been to become a chess grandmaster. He spent thousands of hours practicing, and even moved from his home in the former Yugoslavia to Russia so that he could train with the world's best teachers. But he eventually plateaued. He couldn't increase his rank beyond the level of expert.

Amidzic decided to investigate whether there was something about the way his brain was wired that was handicapping him. So he trained instead to become a neuroscientist, and started to research whether he could identify differences in brain activity between amateurs and grandmasters.

To demonstrate his findings, he got me to play a game of chess against one of Britain's grandmasters, Stuart Conquest, and had us both wired up via a magnetoencephalogram to demonstrate the difference in brain activity. I certainly don't rank anywhere near grandmaster or even expert, but I can think logically and analyze a chess position to see what the best next move might be.

I lost the game quite quickly. But that wasn't the result that interested me. It was the results of the magnetoencephalogram that were striking. It turned out that we were using very different bits of our brain to play. I seemed to be using up more brain activity but achieving less in the way of results.

Amidzic's research has revealed that an amateur player like me is using the medial temporal lobe, which is at the center of the brain. This is consistent with the interpretation that the amateur's mental acuity is focused on analyzing unusual new moves during the game. This is like a more articulated conscious analysis of what the consequences of each move would be, and it could probably be expressed out loud by the amateur player to provide a commentary on their thought process.

In contrast, the grandmaster was tapping into activity in the frontal and parietal cortices and bypassing completely the medial temporal lobe. The frontal and parietal cortices are areas of the brain that are more often associated with intuition. Using these areas involves accessing long-term memory and is involved with a more subconscious thought process. A grandmaster might sense that a move is a good one, even if they can't articulate why. The brain doesn't work hard producing a logical rationale for the feeling, wasting energy in the medial temporal lobe. It was shortcutting conscious thought to arrive at a solution. It's as though my brain is running around like a mad gazelle, but the grandmaster's brain sits there like a lion, wasting no excess energy before making the killing move.

Controversially, Amidzic believes that by scanning an amateur player you can tell whether they have the brain to potentially become a grandmaster, because even at the outset of their career they are already accessing the frontal and parietal cortices as they play.

"Everyone wants to think that you can achieve, that you can be what you want to be," Amidzic said, "and if they are not able to achieve it in life, then you have someone responsible for this, their mother or the government or father's support . . . lack of money or whatsoever, so they have some explanations."

But Amidzic believes that fundamentally, success depends not on investing 10,000 hours of practice, or on having access to great teaching and education, but rather on genetics. "You are born a grandmaster or you are born an average chess player or you are born a great mathematician or musician or soccer player, whatever. People are born, not created. I just don't believe, and I don't see any evidence whatsoever, that you can make or create a genius."

He recalls scanning a child whose father was desperate for him to become a grandmaster. He could tell that the child's brain was stuck analyzing things in the medial temporal lobe. He believed that this kid was never going to push his rating beyond expert and advised the father to think about another pursuit. Apparently the father ignored the advice, and Amidzic has subsequently been proved right in his assessment.

The key is finding an activity for which the brain seems to have a good intuition. In Amidzic's case, he believes that in the end it was neuroscience, not chess, that he was predisposed to excel in: "Life is funny. I'm more famous for this than I would have been as a chess player."

After my brain activity was analyzed while I was playing chess, I learned that I would probably never make the cut as a grandmaster. My brain wasn't finding the shortcuts to good moves but was getting bogged down in taking the long road through my medial temporal lobe. However, Amidzic suggested that if we scanned my brain while I was doing math, then indeed we would see that I was accessing this intuitive part of my brain.

It's not clear from his research whether this is truly all the product of genetics or whether you can train a brain. But his research does seem to have identified that a brain that is engaged in performing at its peak takes advantage of shortcuts to avoid having too much thought cluttering the journey to a solution.

CHAPTER 10

THE IMPOSSIBLE SHORTCUT

Puzzle: At the Glastonbury Festival I often perform
a gig in the Astrolabe Theatre. Afterward I try to
visit all the other stages. Can you find me the
shortest path that starts and finishes at the
Astrolabe and visits all the other stages on
the map once and once only?

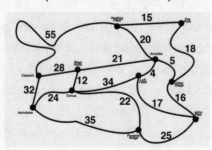

Figure 10.1. Map of Glastonbury Festival

NOT EVERY PROBLEM HAS a shortcut. We have seen how any challenge that demands a physical change to the body, such as learning a musical instrument, rewiring the brain through therapy, or training to be an athlete, necessitates time and effort to achieve. But it turns out that there might be another range of challenges that have no shortcuts. Mathematicians now believe that there is a whole slew of problems that cannot be solved without doing the hard slog of checking all the possible solutions.

Are you a teacher trying to plan next year's class schedule? A hauler planning the best route for your fleet of trucks to deliver goods? A supermarket stocker trying to find an efficient way to put boxes on the shelves? A soccer fan wanting to know if your team can still come out on top of the league? A sudoku fan in search of a good strategy to solve those fiendish puzzles? In all these cases, you're looking for shortcuts. But alas, these might be challenges where better thinking won't help to find a solution. Even Gauss will have to do the hard work of checking all the possible scenarios to find the solution. Perhaps the most striking thing is that mathematics, the art of the shortcut, is out to prove that for certain problems there are no shortcuts.

The classic problem that mathematicians believe has no shortcut to solve it is called the Traveling Salesman Problem. This is the challenge of finding the shortest path around a network of cities. The name seems to have its origins in a handbook for traveling salesmen published in 1832, where the problem is first formulated together with some example tours through Germany and Switzerland. To date, mathematicians have come up with nothing cleverer than trying all the different possible paths to guarantee finding the shortest.

The trouble is that the number of possible routes escalates as I add more cities, and testing every possible route becomes totally impracticable even when implemented on a computer. Surely there must be a quicker way to pick out the solution. Couldn't an Euler or a Gauss or a Newton find some clever strategy to sniff out the shortest route? What about, for example, just always choosing the city that's closest to the one that you are currently visiting? This is called the Nearest Neighbor algorithm. Often this can produce a pretty good route, one that is only 25 percent longer than the optimal path. But it is quite easy to rig up networks where the algorithm ends up producing the longest path through the cities rather than the shortest.

Some algorithms have been developed that do guarantee a route that is never more than 50 percent longer than the optimal path, whatever network the algorithm is given. But what I'm looking for is a clever shortcut that will sniff out the best route without an exhaustive search. This problem has so vexed mathematicians that many have begun to suspect that no such shortcut exists. Indeed, one of the seven Millennium Prize Problems, chosen at the beginning of the twenty-first century as the greatest unsolved mathematical problems, is the challenge to prove that there is no shortcut to solving the Traveling Salesman Problem. The problem is called PvNP, and the mathematician who can prove there is no shortcut will earn themselves a million-dollar prize as a reward.

What Is a Shortcut?

To win the million-dollar prize, it is important to actually define mathematically what would constitute a shortcut in this context. The difference between the long way and the shortcut mathematically translates into an algorithm that takes an exponential amount of time to reach a solution versus an algorithm that only needs polynomial time. What exactly do I mean by this?

The task at the heart of this challenge is to come up not just with a method that will work on one puzzle but with an algorithm that can work whatever version or size of the puzzle it encounters. The question is how long the algorithm takes depending on the size of the puzzle I give it. For example, suppose I have a set of 9 tiles with different patterns on each. I want to arrange the 9 tiles in a 3 × 3 grid such that the patterns along the sides match the patterns on adjacent squares.

Figure 10.2. Nine tiles arranged so patterns along the sides
match the patterns on adjacent squares

How many different ways are there to arrange the tiles? I've got 9 choices for the tile I put in the top left corner of the grid. This tile can be arranged in 4 different orientations. That's a total of 9 × 4 = 36 different choices. The next position along has a choice of 8 remaining tiles, each of which has 4 different orientations again. As I go through the grid I find that the total number of ways of arranging all these tiles is:

$$9! \times 4^9$$

If a computer can perform 100 million checks in a second, this would take just over fifteen minutes. Not too bad. The point is how quickly the time requirement escalates if I increase the number of tiles. What if I consider now 16 tiles in a 4 × 4 grid? Using the same analysis, the number of combinations to check is:

$$16! \times 4^{16}$$

This ramps up the amount of time to check them all to 28.5 million years. Move to a simple 5 × 5 grid and I'm well beyond the lifetime of the universe, which is a mere 13.8 billion years.

Given a grid with n locations, the number of arrangements is given by $n! \times 4^n$. The number 4^n is what is known as a function that grows exponentially with n. I explained the dangerous way this function escalates in Chapter 1, when the king of India had to pay for the game of chess with grains of rice that grew in number exponentially across the board. The factorial $n!$ is in fact a function that grows even faster than exponentially.

This is the mathematical definition of the long way round. Any algorithm to solve a problem where the time it takes to calculate the solution ramps up in this exponential manner as the problem increases in size is going to be the sort of problem for which I would like to find a shortcut. But what will qualify as a good shortcut? This is the discovery of an algorithm that remains relatively fast to find a solution as I increase the size of the problem—so-called polynomial time algorithms.

For example, suppose I have a random selection of words and I want to put them in alphabetical order. How long would this take as the list of words gets longer and longer? A simple algorithm to do this looks through the original list of N words and then pulls out the word that is before all others in the dictionary. Once I have done that I just do the same again to the $N - 1$ words left. In this way I need to scan $N + (N - 1) + (N - 2) + \ldots + 1$ words to sort them. But thanks to Gauss's classroom shortcut I know that this takes $N \times (N + 1) / 2 = (N^2 + N) / 2$ scans in total.

This is an example of a polynomial time algorithm because as N, the number of words, increases, then the number of scans needed just goes up as a quadratic equation in N (squaring N). What I am after in the case of the Traveling Salesman Problem is an algorithm that, given N cities I must make my way around, can find the shortest path by checking, say, only N^2 or a quadratic number of routes.

The first algorithm that we cook up is, unfortunately, not polynomial. Essentially, I first choose one city to visit, then choose the next city, and so on. This will mean checking $N!$ routes given a map with N cities. As I mentioned earlier, that is worse than exponential. The challenge is finding a better strategy than testing all routes.

A Shortcut to a Shortcut

To show that it's not impossible that such an algorithm might exist, consider a problem that at first sight seems equally intractable. Choose two locations on the traveling salesman's map. What is the shortest path between these two cities? At first sight it looks like there are still a lot of different options to consider. After all, I could begin by visiting any of the cities joined to our starting city, and then going to one of the cities joined to that city. I seem to be cranking up to something that is again exponential in the number of cities.

But in 1956 Dutch computer scientist Edsger W. Dijkstra came up with a much cleverer strategy that would find the shortest path between two cities in the same quadratic time that it took to sort the words into alphabetical order. He'd been contemplating the practical problem of what the fastest route was between the two Dutch cities of Rotterdam and Groningen.

> One morning I was shopping in Amsterdam with my young fiancée, and tired, we sat down on the café terrace to drink a cup of coffee and I was just thinking about whether I could do this, and I then designed the algorithm for the shortest path. It was a twenty-minute invention. . . . One of the reasons that it is so nice was that I designed it without pencil and paper. I learned later that one of the advantages of designing without pencil and paper is that you are almost forced to avoid all avoidable complexities. Eventually that algorithm became, to my great amazement, one of the cornerstones of my fame.

Consider the following map:

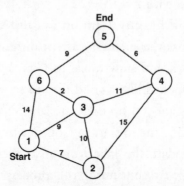

Figure 10.3. What's the shortest path between city 1 and city 5?

In Dijkstra's algorithm I will set out from the starting city, City 1. At each stage of my journey I am going to attach a running total to the cities on the way to help me to find the shortest route. First I label all the cities that are connected to the starting city with the distances to get to them. In this case City 2, City 3, and City 6 get labeled 7, 9, and 14, respectively. My first move will then be to move to the nearest of these cities. But be warned. Once the algorithm has worked its magic to solve the problem, this might not end up being the best city to move to first.

In this picture I begin by moving to City 2, since it is the shortest distance away from my starting point, City 1.

I then mark City 1 that I've just left as "visited." From my new city, City 2, I am going to consider updating the labels on all the cities connected to it. So in this case I'll possibly update the labels on City 3 and City 4. I start by calculating their distances from the start city, City 1, that go via my current city, City 2. If this new distance is shorter than the current label on that city, then I update the label with this new distance. If the distance is longer, then I keep the previous label. In the case of City 3, the new distance (7 + 10) is longer, so I keep the original label, 9. Sometimes the city won't yet have a label, like City 4, because it wasn't connected to the previous cities. I label this new city I've reached with the distance I've just calculated to get to it. So in this case City 4 is labeled with 7 + 15 = 22.

I then again mark the city I am on as "visited" and move to the unvisited city that has the smallest current distance from my starting position. In the example I now move to City 3. Here is an example where, although it looked like a good idea to set off from my starting point for City 2, distances from there are large. So the algorithm is already probably guiding me to start my journey by going to City 3.

Once again I update the labels on any cities unvisited that are connected to City 3. By continuing this process I will eventually get to the destination, City 5, and it will have a label on it that represents the shortest distance from the city I started from. I can then trace back through the journeys to see which cities I went through to get there with this distance. In my example notice it wasn't via City 2 in the end.

How long does this take me in terms of steps I need to implement to find the shortest route? With N cities, it is rather similar to putting words in alphabetical order. Each step removes a city that no longer needs to be considered. So the run time of the algorithm is N^2 or quadratic in N. A shortcut in our mathematical language!

But a shortcut in this mathematical language could still take a very long time to come up with the answer in reality. Mathematicians generally regard polynomial time algorithms as the shortcuts we are looking for. Quadratic algorithms are pretty fast. But even though cubic, quartic, and quintic polynomials are regarded as mathematically fast, they can nonetheless still physically take a long time.

If a computer can do 100 million actions per second, then for small N, I won't have too much of a problem. But there is a massive difference in time between an algorithm that finds the answer in N^2 steps and the algorithm that takes N^5 steps. In one second the N^2 algorithm can check a network with 10,000 cities. The N^5 algorithm would need 31,710 years to check the same number of cities! Yet that's still regarded as a mathematical shortcut. It's certainly a shortcut compared to the exponential algorithm we currently have, which would require way beyond the lifetime of the universe to check a network with 10,000 cities. Indeed, even a calculation with 100 cities is beyond the lifetime of the universe for a 2^N exponential algorithm.

For practical purposes it is still worth battling to find algorithms that require the fewest powers of N to run through. Some shortcuts are shorter than others.

Needles in Haystacks

You might hope that if you aren't a traveling salesman, then you won't be affected by the lack of a shortcut to find the shortest path to get you to all your customers. The trouble is that there are many problems out there that share the same complexity issues. For example, in engineering you might want to wire a circuit board that has 100 different locations, and you need to find the most efficient way to get your robot to lay down the wires to make the circuit. Given that the robot is going to be knocking out thousands of these things each day, just a few seconds' decrease in the time it takes for the robot to travel around the network could save the company a huge sum of money. But it isn't just traveling around a network that we'd love to find shortcuts for. Here is a selection of challenges that have the same quality as the Traveling Salesman Problem, problems that we believe may have no shortcuts to find a solution. It may simply be impossible to avoid the long slog, even for the great Gauss!

The car trunk challenge. You have boxes of different dimensions that you want to transport in the trunk of your car. The challenge is finding the selection of boxes that wastes the least amount of space. It turns out that there is no clever algorithm that will take in the dimensions of the boxes and spit out the best combination. For example, suppose all the boxes are of the same height and width, exactly the same as the internal dimensions of your trunk. But the boxes are of different lengths. Your trunk is 150 cm long, and the boxes available for packing have the following lengths: 16, 27, 37, 42, 52, 59, 65, and 95 cm. Is there a clever way to pick the combination of boxes that will fill the trunk as efficiently as possible?

The school schedule challenge. Every school faces the challenge at the beginning of the school year of preparing a class schedule. But the choices that students make place restrictions on what can be scheduled when. Ada has chosen to do chemistry and music, so these classes can't be scheduled at the same time, while Alan has chosen chemistry and film studies. But there are only eight periods in the day. The school has got to find some way of fitting them all in without an overlap. Given these sorts of constraints, putting together a class schedule can sometimes feel like laying a carpet that doesn't quite fit the room. As soon as you seem to have got it fitting in one corner, you find the carpet pops up in the other corner of the room. Or it's like doing a sudoku where you think you've got the solution, only to find two 2s in the same row. Argghhh!

Sudoku. If you have tried solving some of the fiendish versions of this Japanese puzzle, then you will often hit points where you just seem to have to make a guess at the next number, followed by working out the logical implications of this guess. If it turns out to be a wrong guess and you get a contradiction, then you are forced to backtrack to the moment you made your guess and to try out a different pathway.

The dinner party challenge. A challenge that's similar to the school schedule arises if you have friends you'd like to invite to dinner but there are certain friends of yours who just don't get along, so you can't invite them to the same dinner. Trying to find the minimum number of dinner parties you'll need to host in order to ensure that everyone comes to dinner but no one comes to a dinner with someone they don't like is similar to the challenge of putting together the school's class schedule.

Map coloring. If you take any map and try to color countries so that no two countries with a common border have the same color, then it is always possible to achieve this with four colors. But can you get away with just three colors? Again, the only algorithm that we have to determine if three colors might suffice is by trying to go through all the different ways to color maps. Just like the sudoku, you can start coloring and then a choice made early on forces two countries to have the same color. Trying out all the different possibilities is exponential in the number of countries.

The fact that you only need at most four colors was one of the great theorems proved in the twentieth century. It had been shown in 1890 that five colors will do. The proof was not too complex. It relied on a shortcut that mathematicians often use. Suppose there are maps that can't be colored with five colors. Take an example with the smallest number of countries. Then by some clever analysis one can show how to remove one country and still the resulting map can't be colored with five colors. But this contradicts the fact that the map you started with was meant to be one with the smallest number of countries.

The frustrating thing was that this clever shortcut didn't seem to work if one tried to show that four colors suffice. Mathematicians couldn't show why removing a country would still give you a map that couldn't be colored. And yet no one could come up with a counter-example.

Eventually, in 1976, a proof was found that four colors suffice. But it certainly wasn't regarded as a shortcut. Indeed, the proof required the brute force of a computer to check thousands of cases that it was impossible for a human to sort through. This proof was a turning point in mathematics: the first time that a computer was used to force through a path to the end of the proof. It was like we'd hit a mountain range and found no clever path through to the valley on the other side. Instead we used a machine to bore a hole through the mountainside.

There were many in the mathematical community who felt a certain unease at the way a computer was used to prove this theorem. Proofs were meant to give humans an understanding of why four colors suffice, not simply determine whether it was true. The human brain is limited in the connections it can make, which is why the shortcut is absolutely essential for the brain to feel like it gets why something is the way it is. If the proof is forced to take the long way around, then it's as if the proof can't be loaded into the brain, and it feels like we're being denied a true sense of understanding.

A problem related to the map-coloring challenge is to take a network consisting of points joined together with a selection of lines. The lines are like the borders between the countries. The challenge is to know what the fewest number of colors is needed to color the points so that no two points have the same color if they are connected by a line.

Soccer. I think one of my favorite examples of a problem that we can't find a shortcut to solve relates to soccer—not the playing of it, but that wonderful challenge that starts to rear its head toward the end of a season: Is it still mathematically possible for my team to win the championship given where it currently stands in the league? You might think this is a simple task—I just make sure that my team wins all its games, getting 3 points for each win, and then check if that is enough to come out on top. The point is that it's all the other games between teams that I need to worry about. Clearly I want to make the team that is currently at the top of the standings lose lots of games. But that will have the effect of the teams they are playing winning and getting more points. What if I end up assigning them too many points and they come out on top?

This is another problem where I've got to consider so many different combinations of matches and outcomes. As I assign wins, losses, and draws, time and again I find myself backtracking, as in a sudoku, because one of the outcomes I've assigned has messed up my careful balancing act.

The only algorithm that exists at present is exponential in the number of teams in the division. The challenge is to find a shortcut that will tell me quickly whether my team still has a mathematical chance of winning the league.

But why I like this problem so much is that in the past, when I was in school, there *was* such an algorithm. What's happened in the interim? It's not that we lost the algorithm; rather, the way points are assigned has changed. When I was in school a team got only 2 points for a win, 0 points for a loss, and 1 point if they drew. This was felt to encourage teams to go for boring draws. So in 1981 a decision was made to try to incentivize teams to push for the win. Instead of 2 points, teams could earn 3 points for a win. It seems like an innocuous change, but it had a dramatic effect on the challenge of working out whether a team could still come out on top in the league standings.

The crucial change is that before 1981, the total number of points shared out across the teams didn't depend on who won, lost, or drew. There are 20 teams that play each other twice, home and away, which means 20×19 matches. Each match in the old system had 2 points that were distributed according to the result, which means that the total number of points at the end of the season was $2 \times 20 \times 19 = 760$ points shared among the 20 teams.

But today things are very different. Each match could result in 3 points being awarded to the winner or just 1 point each if the match is a draw. If all matches across the season were drawn, that would mean a total of 760 points again. But if there are no draws, then the total is

$3 \times 20 \times 19 = 1140$ points. This new variance in the total number of points means that the previously successful algorithm that could tell me if my team still had a mathematical chance of winning the league no longer works.

The fascinating thing with all these problems is that if you happen to stumble on a potential solution, it is quick to check whether it really solves the challenge. I like to call these needle-in-a-haystack problems: the initial challenge of finding the needle requires a long, exhaustive search with little to help you home in on where the needle is, but as soon as your hand lands on the needle, you know it! Cracking a safe can take a long time, trying one combination after another, but once you put in the right combination the door opens immediately.

There is a rather extraordinary feature of these needle-in-a-haystack problems—or, to give them their technical name, NP-complete problems. You might expect that each problem is going to require its own customized strategy to try to discover an algorithm that will find the solution in as short a time as possible. But it turns out that if you ever did discover a fast polynomial time algorithm that finds the shortest path in any map the traveling salesman might be faced with, that means there must be such algorithms for every other such problem. This at least shortcuts the challenge of finding shortcuts. If there is a shortcut to one problem, it can be converted into a shortcut for any other challenge on our list. To misquote Tolkien: one shortcut to solve them all.

I can give a hint of why this might be true by seeing how it is possible to translate some of the problems I have described into each other. Take, for example, the school schedule problem. I had classes and time slots and clashes between classes that I needed to avoid. Using this information, I can build a network where each class is a point in the network, and the clashes between classes will correspond to lines that I draw between two classes that clash. Then assigning time slots becomes exactly the same challenge as coloring the points in the graph such that no two points with a line joining them have the same color.

Exploiting the Lack of Shortcuts

There are some settings where it is quite important that there is no shortcut. This is the case when it comes to making unbreakable codes. Code-makers want to exploit the fact that there doesn't seem to be any way to crack a coded message beyond an exhaustive search of possibilities. Take, for example, a combination lock. A lock that has 4 dials with 10 numbers on each dial is going to require checking 10,000 different numbers, from 0000 to 9999. Sometimes a badly manufactured lock can give away the positions where the lock will release because there is a physical shift in the locking device when the first dial is set in place but in general there is nothing a thief can do to shortcut trying all the combinations.

But other cryptosystems have revealed weaknesses that can be exploited to create shortcuts. Take the classic Caesar cipher or substitution cipher. This is a code that systematically exchanges letters for alternative letters. For example, every instance of A gets replaced by the letter G. Then B is replaced by one of the remaining letters. In this way each letter of the alphabet gets assigned a new letter. There are a lot of different codes one can choose. There are 26! different ways to rearrange the letters of the alphabet. (In some of the rearrangements a letter might remain the same, X encoded by X, for example. An interesting challenge: How many codes are there where all the letters get changed?) To give a sense of the size of this number, 26! seconds translates into more time than there's been since the Big Bang.

If a hacker intercepts a coded message, they have to try a lot of different combinations to decode the message. But there is a weakness to this code, which the ninth-century polymath Ya'qub al-Kindi spotted: some letters appear more frequently than others. For example, in the English language E is the most common letter to appear in any text, occurring 13 percent of the time followed by the letter T, which crops up 9 percent of the time. Letters also have their own personalities, which get reflected in other letters they like to be associated with. In English, Q is invariably followed by U, for example.

Al-Kindi realized that a hacker could exploit this as a shortcut to attacking a message encoded with a substitution cipher. By doing a frequency analysis on the encoded text and matching the most frequent letters to those that occur most frequently in plain text, the hacker starts to get the first inroads into unraveling the message. It turns out that using a frequency analysis is an amazing shortcut to cracking these codes, which are far less secure than they first appear to be.

The Germans during World War II thought they'd found a different way to use a substitution cipher that could avoid this shortcut to cracking messages. Their idea was to use a different substitution cipher to encode each new letter in the message. This meant that if they were encoding *EEEE*, then each *E* would be sent to a different letter, which would block any attack using al-Kindi's frequency analysis. The way they managed to do this multiple-substitution cipher encoding was to build a machine: the Enigma machine.

You can still see one of these machines on display at Bletchley Park, the home of the United Kingdom's codebreakers during the war. At first sight the machine looks like a conventional typewriter with a keyboard, but there is a second copy of letters running above the keyboard. When I pressed one of the keys, this caused one of the letters above the keyboard to light up. This was how I would encode this letter. Essentially the wires in the machine were scrambling the letters in a classic substitution cipher. But at the same time as I pressed the key, I could also hear a clicking and saw one of three rotors sitting at the heart of the machine click over one step. When I pressed the same letter again, a different bulb lit up. This is because the wiring from the keyboard to the lights had been rearranged. The wires are connected via the cogs, and when the cogs change alignment, so does the wiring in the machine. In this way the clicking over of the cogs ensures that the machine uses a different substitution cipher for each letter it is encoding.

The whole thing looks uncrackable. There are 6 different cogs I could use to set up the machine, and each cog could be started in 26 different settings. Plus there was a whole set of wires at the back of the machine that could add another fixed level of scrambling. It meant that there were 158 million million million different ways that the machine could be set up. Trying to find out which one an operator had used to encode a message looked like the ultimate task of looking for a needle in a haystack. The Germans were utterly confident that the code this machine produced was uncrackable.

But they hadn't reckoned on the clever skills of mathematician Alan Turing, a twentieth-century Gauss, who, sitting in Bletchley Park, sniffed out a weakness in the system that could be exploited to shortcut an exhaustive search. The key was that the machine never encoded a letter by the same letter. The wiring would always send the letter to a different letter. It looks like an innocent fact about the machine. But Turing saw how this could be used to chase around the machine and squeeze out a much more limited set of possibilities for how a particular message had been encoded.

He still needed to employ a machine to do the final search. The huts at Bletchley Park would hum all night to the sound of the Bombes, the name the team gave to the machines that implemented Turing's shortcut. But each night it would give the Allies access to the messages the Germans thought they were sending securely.

Prime Suspects

Today the codes that protect our credit cards as they fly across the internet are exploiting mathematical problems that we believe by their nature have no shortcuts. One of these ciphers, called RSA, relies on those enigmatic numbers, the primes. Each website secretly chooses two prime numbers of about 100 digits each in length, which are multiplied together. The result is a number with roughly 200 digits, which is then made publicly available on the website. This is the website's code number. When I visit a website, my computer receives this 200-digit number, which is then used to do a mathematical calculation involving my credit card. This scrambled number is sent across the internet. It is secure because to undo the calculation a hacker has to find two prime numbers that multiplied together give the website's 200-digit code number. The reason this cryptography is regarded as secure is that this seems to be a needle-in-a-haystack problem. The only way mathematicians know to find these primes is to try one after another, hoping that we suddenly hit on the needle where the number exactly divides the website's code number.

Gauss himself wrote about the challenge of cracking numbers into primes in his great treatise on number theory, *Disquisitiones Arithmeticae*.

> The problem of distinguishing prime numbers from composite numbers and of resolving the latter into their prime factors is known to be one of the most important and useful in arithmetic. It has engaged the industry and wisdom of ancient and modern geometers to such an extent that it would be superfluous to discuss the problem at length. . . . Further, the dignity of the science itself seems to require that every possible means be explored for the solution of a problem so elegant and so celebrated.

Little did he realize how important this problem would become in the age of the internet and e-commerce. No one to date has found a shortcut for finding the primes that divide large numbers, not even the great Gauss himself. The only way we know is to check one prime after another. The number of primes that one needs to check to crack a 200-digit number is so vast that such an attack is totally ineffective. Currently we don't know of a shortcut to avoid an exhaustive search. We think that the challenge of factoring might be intrinsically difficult. That is one of the open problems that mathematicians are currently working on. Can we prove that there is no shortcut to find the primes?

But hold on. How does the website decode messages? The point is that it started the process by choosing the two primes of roughly 100 digits each, which were then multiplied together to generate its 200-digit public code number. The website is the only one in possession of the primes that will undo the calculation.

But isn't finding primes one of those problems that mathematicians haven't solved yet? Cracking the secret to how the primes are arranged through the universe of numbers, called the Riemann hypothesis, is another of the seven Millennium Prize Problems. But even though mathematicians don't really understand exactly how the primes are distributed, we do actually have an interesting shortcut to finding big primes for these internet codes. It depends on a discovery that the great seventeenth-century French mathematician Pierre de Fermat made about prime numbers. He proved that if p is a prime number and I take any number n less than p, then raising n to the power p will have remainder n when I divide by p. For example, $2^5 = 32$, which has remainder 2 on division by 5.

This means that if I want to test whether a candidate number q is prime, if I can find a number less than q that fails this test, then I know q isn't prime. For example, $2^6 = 64$, which has remainder 4, not 2, on division by 6. This means 6 can't be prime because it fails Fermat's test. This would not be a very useful test if, say, only one number less than q fails the test. It would mean testing potentially all numbers less than q, and then I might as well just check indivisibility directly. The great advantage of this test is that if a potential prime number fails, it fails spectacularly. Using Fermat's ruse, over half the numbers less than q will bear witness to q's failure to be prime.

There is a fly in the ointment, however. There are some numbers that behave like primes, in that there are no Fermat witnesses betraying them, but which nonetheless are not prime. They are called pseudoprimes. But in the late 1980s two mathematicians, Gary Miller and Michael Rabin, were able to refine Fermat's approach to produce a guaranteed test for primality that would run in polynomial time. The only caveat was that in order to make this amazing shortcut to primality, the two mathematicians had to assume that they could first climb to the top of a very tall mountain: the Riemann hypothesis (or a generalization of the conjecture).

Miller and Rabin could prove that, provided mathematicians found a way to clear this summit, they could guarantee a shortcut on the other side to finding primes. This summit is so important partly because so many mathematicians have shown that it would give access to a whole slew of shortcuts. I myself have several theorems that show why something is true provided I can first prove the Riemann hypothesis is true.

But one should never give up on the chance that there might be a sneakier way around the mountain. In 2002 the mathematical community was rocked by the exciting news that three Indian mathematicians, Manindra Agrawal, Neeraj Kayal, and Nitin Saxena, based at the Indian Institute of Technology in Kanpur, had come up with a way to test whether or not a number was prime in polynomial time without having to go over Mount Riemann. The remarkable thing is that the second two authors of the discovery were undergraduate students working with Agrawal. Even Agrawal, the senior member of the team, was unknown to most number theorists in the mathematical community. It reminded many of the story of the great Ramanujan, who exploded onto the mathematical scene in the early twentieth century after writing to Cambridge mathematician G. H. Hardy about his mathematical discoveries.

Although the breakthrough established a test for primality that worked in polynomial time without assuming that you could cross Mount Riemann, it was not an algorithm that was terribly practical in real terms. As I mentioned earlier, it's important to know the degree of the polynomial. If it's quadratic, then things will run fast. The original algorithm proposed by Agrawal, Kayal, and Saxena had a polynomial with degree 12. This was cut down to 6 by American mathematician Carl Pomerance and Dutch mathematician Hendrik Lenstra, but as I explained earlier, although mathematically a shortcut, practically this slows down quite quickly.

Given that internet security depends on a healthy supply of big primes, how do websites find them quickly enough to run financial services efficiently? The key is to use an algorithm that doesn't have to guarantee a number is prime but gives the website a high degree of confidence it is.

Remember that if a number is not a prime or a pseudo-prime, then half the numbers less than it will fail the Fermat test. But what if I was really unlucky and tested the half that passed the test? To guarantee finding a witness to the non-primality seems to require testing half the numbers. But what are the chances of missing a witness? Say I did 100 tests and didn't get a witness. Either that means the number is a prime or a pseudo-prime or there is a one in 2^{100} chance that I missed all the witnesses. That is a bet I'd be prepared to take! Those are very long odds that the number is not a pseudo-prime but I failed to find a witness.

Although we have great algorithms, both deterministic and probabilistic, for finding primes to make these codes, it seems that conventional algorithms don't exist for breaking codes. What about something a little more unconventional?

Quantum Shortcuts

One of the problems that conventional computers face when trying to do something like factor numbers is that they are required to carry out one trial before they can move on to the next test. What I really want is to divide the computer up into bits and get each bit to do a different test. Parallel processing is a very effective way to speed up actions. Take the case of building a house. In a competition held in Los Angeles to see which team of builders could erect a house the fastest, it was a team with two hundred builders working in parallel who won the race, building a house in four hours. There are obviously some tasks that rely on things being done in sequence. Building a high-rise or digging an underground parking garage relies on each floor being built before the next one is added to it. But testing numbers to see if they divide the number I am trying to factor is a perfect task for doing in parallel. Each task does not rely on the outcome of any other check.

The trouble with parallel processing is that I still have a physical capacity problem. By dividing the problem in half I have cut down the time to implement the check by 2, but I have doubled the amount of hardware I need. This approach doesn't really address the issue of factoring.

But what if I could do this parallel processing without the need to double the hardware? This was the idea of Peter Shor, a mathematician working at Bell Labs in the 1990s, who realized that one could exploit some rather unconventional computing to test things simultaneously. The idea was to draw on the bizarre physics of the quantum world. In quantum physics, it is possible to set up a particle, such as an electron, so that before it is observed it essentially is located in two positions simultaneously. Let's call these two positions 0 and 1. This is called being in quantum superposition. The advantage of this is that the hardware hasn't doubled: it's just one electron. But this electron is actually storing two pieces of information, not just one. This is called a qubit. Instead of a conventional computer, which has to set a bit in either the on position or the off position, a 0 or a 1, this qubit splits into parallel quantum worlds, one in which the switch is set to 0, the other where it is in the 1 position.

The idea would therefore be to string a whole load of these qubits together. For example, if I were able to put 64 qubits in quantum superposition together, then this bank of 64 qubits could simultaneously represent all numbers from 0 to $2^{64} - 1$. A conventional computer would have to sequentially run through all these numbers, putting each bit into the 0 position or the 1 position. But the quantum computer can do it simultaneously. It's as if my conventional computer, like the electron, is suddenly living simultaneously in parallel universes. In each one the 64 qubits are set to a different number.

But now comes the really neat bit. In each parallel world, I get the computer to check if this number that it is representing divides our crypto number. But how can I make sure that the quantum computer can pick out the one world where factoring happens? This was Shor's brilliant trick that he built into his quantum algorithm. When I observe a quantum superposition, it has to make up its mind and collapse into one state or the other. Essentially it chooses either the 0 position or the 1 position. There are probabilities that determine which way it will go.

What Shor managed to do was to rig up an algorithm in which, after doing the test factoring in each parallel universe, the probability would be overwhelming that it would collapse in the world where factoring happened. All the other worlds where this failed were sufficiently similar that they all canceled each other out; the only one that stuck out was the world where factoring happened.

Imagine twelve different directions pointing to the positions on a clock face. If all of them are equal in length, then if I add all these directions together, they all cancel each other and bring me to the center of the clock face. But what if one of them was twice as long as all the others? Now I would be left pointing in the direction that is longer. This is essentially what happens in the quantum observation of the test factorings.

Although Shor wrote the software as far back as 1994, building a quantum computer that could implement the algorithm seemed a distant dream. One of the problems with quantum states is something called decoherence. The 64 qubits start observing each other, and the thing collapses before it can be made to do the calculation. It is one reason that we think that Schrödinger's cat might not be possible. Sure, an electron can be put into superposition, but how can all the atoms that make up a cat be put into a state of being dead and alive at the same time? The large number of atoms start interacting, and decoherence means that the superposition collapses.

But in recent years there has been some amazing progress in isolating simultaneous quantum states. In October 2019, *Nature* published a paper by researchers at Google entitled "Quantum Supremacy Using a Programmable Superconducting Processor." In the paper the team reports being able to rig up 53 qubits in superposition, giving them the ability to represent simultaneously numbers up to about 10^{16}. The computer was able to perform a highly specific task that would have taken a conventional computer 10,000 years to execute.

Although this was very exciting news, the task that the quantum computer was asked to carry out was not on a par with factoring numbers and was very much tuned to the hardware that was being used. Many felt that Google was hyping the "quantum supremacy" headline a bit too much. The IBM quantum computing team was pretty scathing about the announcement and illustrated how the task the Google team was implementing could be done on a conventional computer in days, not 10,000 years. Still, it was a very fascinating result. But it seems like creating a quantum computer that could hack your credit card details is still some way off.

Biological Computing

What about the Traveling Salesman Problem? Could I use unconventional means to find a shortcut? There is a problem related to the Traveling Salesman Problem that researchers have solved using interesting ways to exploit non-conventional computing modalities. Called the Hamiltonian Path Problem, the challenge is to find your way around a network of one-way streets connecting cities on a map.

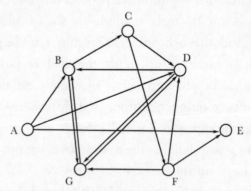

Figure 10.4. Hamiltonian Path Problem: get from city A to city E visiting every other city once only

The problem is to discover a path that begins at one city, say City A, and ends at a second city, say City E, but the path must visit every other city on the way once and once only. Is such a path possible? This turns out to be as complex as the Traveling Salesman Problem. But it is again a problem that is ripe for parallel processing. However, rather than exploiting the quantum world, mathematician Leonard Adleman cooked up an interesting use of biology to attack the problem. (Adleman is the A in RSA, the name of the cryptography that exploits primes to keep online transactions secure.)

At a seminar at MIT in 1994, Adleman announced the TT-100, the supercomputer he'd built to attack the Hamiltonian Path Problem. The audience was rather perplexed when he pulled a test tube from his jacket pocket. It turned out that the TT stands for test tube and the 100 for the 100 microliters this small plastic vial contained. The microprocessors that were doing the work inside this test tube were small strands of DNA.

Strands of DNA are made up from four bases, labeled A, T, C, and G. These bases like to bind to each other in pairs, A to T and C to G. If you cook up short single strands of these bases, called oligonucleotides, then they will try to find another strand with bases that pair up. For example, a strand with ACA will try to find a strand with TGT to bind with to make a stable double strand of DNA.

Adleman's idea was to give each city on a map that you were trying to navigate a label consisting of a string of 8 bases. Then if there was a one-way road between the two cities, he would create a strand of DNA with 16 bases such that the first 8 consisted of the code number of the origin city and the second 8 were the complementary string corresponding to the city at which the road ended. If there was a road into City A and a road out, then the two 16-base strands for these roads would join up along the last 8 bases of the road coming in and the first 8 bases for the road coming out of City A.

Any route around the cities along these roads could actually be replicated in strands of DNA that were binding to each other each time a road went in and out of a city. For example, City A could have label ATGTACCA, City B label GGTCCACG, and City C label TCGACCGG. The road from A to B is then represented by

ATGTACCACCAGGTGC

And the road from B to C by

GGTCCACGAGCTGGCC

These two roads would then bind along the last 8 bases of the first route and the first 8 bases of the second route to show that there is a journey you can make from City A to C.

The great thing is that it is possible to order such strands of DNA in huge amounts from commercial labs. Adleman ordered up enough to explore a network with 7 cities and then just filled his test tube with the strands. Then in an act of parallel processing, the strands started binding together to create lots of different pathways through the network. Of course, many of them violated the challenge of visiting cities only once. But Adleman realized the route he was after would be a strand of DNA of length

8 (the origin city) + 6 × 8 (for each road) + 8 (the destination city)

He could filter these out of the solution and then check for those strands that had each city occurring somewhere in the sequence, in a process similar to genetic fingerprinting.

Although the whole process took over a week, it nonetheless opened up an intriguing possibility to exploit the world of biology to create machines that can parallel-process efficiently. Chemists are quite happy to play with a mole's worth of a substance, but a mole contains an Avogadro's number of molecules. This is slightly more than 6×10^{23}—a huge number. Adleman believes that exploiting the very small in the biological world can be a shortcut to navigating the very big in conventional computational challenges.

It may be that Nature has already figured this out. It turns out that a strange organism called slime mold is rather good at finding the most efficient routes to navigate a map. Slime mold (*Physarum polycephalum*) is a plasmodial, single-celled organism that grows outward from a single point, searching for food sources. The food it loves most is oat flakes.

A team in Oxford and Sapporo decided to set their slime mold the challenge of finding the shortest route around oat flakes laid out in the locations of stations on the Tokyo rail network. Human engineers had spent years putting together the most efficient way to connect the cities. How would the slime mold do in comparison?

At the outset, the slime mold knows nothing about the locations of the oat flakes and so starts growing in all directions. But as it begins to encounter a food source, what happens is that the many branches it has sent out that fail to find food die back, leaving only the most efficient route between the food sources. Within hours the slime mold is refining its structure and creating routes between the food sources that efficiently navigate the different locations.

What was remarkable to the team that built the experiment is that the resulting patterns of the slime mold resembled very closely the way humans had laid out the Tokyo rail system. Humans had taken years. The slime mold did it in an afternoon. Does this single-celled slime mold know a shortcut that might help us solve one of mathematics' great unsolved problems?

Solution to the Puzzle

Here is the shortest path around the traveling salesman's map. It took me a long time to check that there wasn't a shorter path.

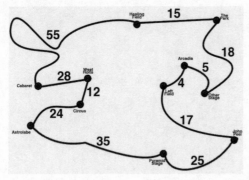

Figure 10.5. The shortest path around Glastonbury Festival:
35 + 25 + 17 + 4 + 5 + 18 + 15 + 55 + 28 + 12 + 24 = 238

SHORTCUT TO THE SHORTCUT

....

Sometimes it's just as important to know when there are no shortcuts to the problem you're trying to crack. Knowing that the long way around is the only way to your destination will prevent you from wasting time in the hope of finding the shortcut. And if you are going to do all the work, then it's worth knowing that you are not wasting your time. You can use the shortcut of changing one problem into a completely different problem to check if the challenge you are trying to crack is actually the Traveling Salesman Problem in disguise. If there are no shortcuts, then perhaps, as the cryptographers have done, this is something that you can exploit to your advantage.

....

ARRIVAL

HUMAN INGENUITY HAS CONJURED up an extraordinary range of different shortcuts, which have accelerated the development of our species over the generations. We would never have arrived at the technologically advanced place we currently occupy without this portfolio of better ways of thinking. Without the shortcut of symbols for numbers, everything beyond 3 just looks like a lot. Our physical journeys across the planet have been made more efficient by understanding its geometry. Although only 566 people have been into space and none farther than the moon, we have used the shortcut of trigonometry to navigate deep into the cosmos.

We have been able to shortcut our journey into the future using the power of pattern recognition and the calculus to get a glimpse of what's coming next before it happens. The shortcut of probability means that we don't have to repeat experiments hundreds of times to understand which outcome is more likely. Rather than wandering aimlessly around the internet looking for what we want, clever ways of analyzing the connections allow us to shortcut our way to our destination. We've even come up with new numbers, like the square root of -1, to create a looking-glass world through which we can step to shortcut our way to a solution. Planes land safely thanks to a trip through this imaginary world.

Sidestepping tedious hard work was certainly the initial reason I set off on my mathematical journey. Avoiding mindless labor appealed to the lazy side of my teenage self. I am grateful to my math teacher, who, rather than pushing the class into tedious repetition and calculation, instead showed me that mathematics was about thinking smart. But looking back, I have also begun to see something of a paradox at the heart of shortcuts.

The job of a mathematician is to discover new ways of thinking smart, but coming up with these shortcuts is not easy. Doing mathematics still necessitates hours of meditation on a problem, thinking and seemingly getting nowhere. And then suddenly there's this rush of understanding, the discovery of the shortcut through the wilds of the problem. But without putting in the hours of meditation and random scribbling on my yellow legal pads, I can't achieve that rush of the discovery of the way through. It's that thrill of the aha moment that I crave. That's the drug. And I get that hit through the discovery of a hidden passage, the shortcut, that gets me through to the other side.

In the end I realized it's not actually because I'm lazy that I dedicated myself to the art of the shortcut. Almost the opposite. It's the hard work of finding a shortcut that makes it so satisfying. Faced with a mountain, I could take a helicopter to the top. I would enjoy the view, but as Robert Macfarlane explained to me, if you are a mountaineer, this defeats the point. It is the satisfaction of achieving the peak that is the reason you put in the hard work—to "walk the flesh transparent."

I remember talking to the Harvard physicist Melissa Franklin about the intellectual challenges of tackling great unsolved problems. At one point she offered me a hypothetical button to press that would give me all the answers to all the problems that I was working on. As I reached out to press it, she grabbed my hand. "Are you sure you want to do that? Doesn't it spoil the fun?"

Natalie Clein expressed the same reservation. If there was a shortcut to playing the cello, maybe it would make performing less attractive. Achieving that ecstatic moment of psychological flow is about combining skill with a difficult challenge.

One of my favorite Hollywood movies is *Good Will Hunting*, partly because it is one of the first places that the Fields Medal, the mathematician's Nobel Prize, gets mentioned in popular culture. But the film also illustrates the importance of spending hours working away at a frustrating problem as the foreplay to the moment when you discover the shortcut that cracks the problem. The central character in *Good Will Hunting* is a janitor in the math department, played by Matt Damon, who sees a problem chalked on the board and immediately realizes how to solve it. The math professors are blown away when they come in next morning to see the solution scrawled on the board. But ultimately Damon's character does not become a mathematician.

For me, the reason is that he finds it too simple. It's the girl he is trying to catch that is the complex problem without any obvious solution, and it's this that motivates his journey at the end of the movie. One of the important traits of the mathematical shortcut is that it should provide a moment of ecstatic release after all the hard grafting of trying to solve the problem head-on.

The shortcuts I'm seeking are not about looking up the answers in the back of the book. That isn't a satisfying shortcut. The best shortcuts are those that emerge after the hard work of struggling with a problem. It almost has a musical quality to it, where the tension of the music is finally resolved.

The paradox that has emerged is that although the motivation for the shortcut might be an initial reluctance to spend ages doing hard work, ultimately I might end up putting in just as much work to find the shortcut. But it is the nature of the curve describing the effort that perhaps reflects why I still enjoy the hard work to the shortcut more.

If I drew a graph of the effort I would expend adding up numbers from 1 to 100, then it would probably look like a constant grind—not varying much over time, the total effort creeping up in a linear fashion. The graph charting the effort to find the shortcut looks much more unpredictable. It has ups and downs. It probably spikes toward the end before swooping down to a low as the shortcut is implemented. The important thing is that from this point on, the graph of effort never has to go above a baseline minimum because after that the shortcut does the work. The long-haul graph is still plowing on at its constant grind.

Another of the curious paradoxes that emerged is one that Hans Ulrich Obrist highlighted. Detours are essential. You often arrive at the best shortcuts by beginning with the detour. The detour that the proof of Fermat's Last Theorem took mathematicians on is worth all the strange highways and byways we encountered on the way. Those detours led to the discovery of many extraordinary shortcuts we were forced to conjure up on our journey.

The power of the shortcut is often that it allows those who follow it the chance to reach their destination quicker. In 2016 the longest and deepest tunnel in the world was opened. The 57 km Gotthard Tunnel runs beneath the Alps joining northern and southern Europe. It took 17 years to build, but the trains that now run through take 17 minutes to get from one end to the other.

One of Gauss's last journeys was to attend the opening of the new railway link between Hanover and Göttingen, but it proved to be his last outing. His health deteriorated slowly, and Gauss died in his sleep early in the morning of February 23, 1855. Gauss asked that one of the discoveries that had inspired his journey to become a mathematician, the geometric construction of a 17-sided figure, be carved onto his gravestone. But when the stonemason tasked with carving the memorial saw the design, he refused to include it. The construction might theoretically produce a 17-sided polygon, but the stonecutter thought it would just look like a big circle.

The shortcuts I spent my time learning as a student took their creators years of deep thought to carve out. But once the tunnel had been created, it allowed those who follow the chance to get to the frontiers of knowledge as quickly as possible. The young Gauss, sitting in his classroom having completed his task of adding the numbers from 1 to 100, had the chance to think about new things. And for me this is the point of the shortcut. If I spend time on mindless work, then I deprive myself of the opportunity for self-exploration, for new discoveries, for expanding my horizons. The shortcut allows me to dedicate my effort to new, exciting, and rewarding ventures.

So I hope the journey we have been on provides you with the short-cuts to think smarter and free up your time for new thoughts. The end of one shortcut is the chance to begin a new journey. Gauss summarized his views on the pursuit of knowledge in a letter to his friend Farkas Bolyai, dated September 2, 1808:

> It is not knowledge, but the act of learning, not possession but the act of getting there, which grants the greatest enjoyment. When I have clarified and exhausted a subject, then I turn away from it, in order to go into darkness again. The never-satisfied man is so strange; if he has completed a structure, then it is not in order to dwell in it peacefully, but in order to begin another. I imagine the world conqueror must feel thus, who, after one kingdom is scarcely conquered, stretches out his arms for others.

A shortcut is not a fast way to finish your journey, but rather a stepping-stone to beginning a new one. It is a pathway cleared, a tunnel dug, a bridge constructed to allow others to quickly reach the frontiers of knowledge so they can make their own journey into the darkness. Equipped with the tools that Gauss and his fellow mathematicians through the ages have honed, stretch out your arms for the next great conquest.

ACKNOWLEDGMENTS

THERE AREN'T MANY SHORTCUTS to the epic task of writing a book, but one of the best shortcuts is to have a great team supporting you. Like the best psychologist, Louise Haines, my editor at Fourth Estate in the United Kingdom, has a wonderful way of asking that probing question that creates an environment in which you, as an author, discover the solutions to the problems you are having. Antony Topping, my agent at Greene and Heaton, has always been another important set of eyes, ensuring I don't get lost down paths that lead nowhere. My copyeditor, Iain Hunt, very patiently wrestled with my mangled English grammar, knocking it into shape.

On the other side of the pond, my US editorial team, Thomas Kelleher and Eric Henney at Basic Books, has done a wonderful job of ensuring my shortcuts take US readers in the right direction. Thanks also to my US copyeditor, Sue Warga, for changing *football* to *soccer* and much more, and to my US agent, Zoë Pagnamenta, for teaming me up with Basic Books.

Each of my contributors to the pit stops that punctuate this book was extremely generous with their time and ideas. I am deeply grateful to Natalie Clein, Brent Hoberman, Ed Cooke, Robert Macfarlane, Kate Raworth, Hans Ulrich Obrist, Conrad Shawcross, Fiona Kennedy, Susie Orbach, Helen Rodriguez, and Ognjen Amidzic for fascinating discussions about their thoughts on shortcuts.

Thank you to the artists Sophia Al Maria, Tracy Emin, Alison Knowles and Yoko Ono for permission to include their *do it* instructions.

305

Writing such a book would be impossible without the time that my professorship provides me. Thank you to Charles Simonyi who endowed the chair and Oxford University for all the support I receive as the Professor for the Public Understanding of Science.

Both Newton and Shakespeare profited from being more productive during times of plague. The writing of this book overlapped with the pandemic that hit the planet at the beginning of 2020. This turned out to be a strange shortcut, as it gutted my diary of distractions, leaving me with time to sit and write. The result was that I finished the manuscript two months before my deadline. My editor, Louise, was in shock when the manuscript arrived. She is used to me being two years late! But it transpired I wasn't the only author to submit early. Indeed, Louise admitted she'd received from her authors novels she hadn't even commissioned. Feedback might take a bit of time, she warned. While I waited, I ended up using lockdown to write a new play—probably a crazy project, given that all the theaters had closed, but I hope it will see the light of day at some point.

A day of writing during lockdown would end with each member of the family emerging from their room to share their day's online adventures. The laughter and love that we shared together of an evening made the hard slog of getting to the finish line of my book that much easier. Thanks to Shani, Tomer, Ina, and Magaly. They were the best shortcut to completing the daunting journey of writing a book.

INDEX